Modernism and ...

Series Editor: Roger Griffin,
University, UK.

The series *Modernism and...* invit
tific and political phenomena to
in modern history and 'modern
ground-breaking specialist mon
impact to expand the application   ....., contested term beyond its conventional remit of art and aesthetics. Our definition of modernism embraces the vast
profusion of creative acts, reforming initiatives, and utopian projects that, since the
late nineteenth century, have sought either to articulate, and so symbolically tran-
scend, the spiritual malaise or decadence of modernity, or to find a radical solution to
it through a movement of spiritual, social, political – even racial – regeneration and
renewal. The ultimate aim is to foster a spirit of transdisciplinary collaboration in
shifting the structural forces that define modern history beyond their conventional
conceptual frameworks.

Titles include:

Marius Turda
MODERNISM AND EUGENICS

Forthcoming titles:

Tamir Bar-On
MODERNISM AND THE EUROPEAN NEW RIGHT

Maria Bucur
MODERNISM AND GENDER

Frances Connelly
MODERNISM AND THE GROTESQUE

Elizabeth Darling
MODERNISM AND DOMESTICITY

Matthew Feldman
MODERNISM AND PROPAGANDA

Claudio Fogu
MODERNISM AND MEDITERRANEANISM

Roger Griffin
MODERNISM AND TERRORISM

Ben Hutchinson
MODERNISM AND STYLE

Carmen Kuhling
MODERNISM AND NEW RELIGIONS

Patricia Leighten
MODERNISM AND ANARCHISM

Thomas Linehan
MODERNISM AND BRITISH SOCIALISM

Gregory Maertz
MODERNISM AND NAZI PAINTING

Paul March-Russell
MODERNISM AND SCIENCE FICTION

Anna Katharina Schaffner
MODERNISM AND EROTICISM

Richard Shorten
MODERNISM AND TOTALITARIANISM

Mihai Spariosu
MODERNISM, EXILE AND UTOPIA

Roy Starrs
MODERNISM AND JAPANESE CULTURE

Erik Tonning
MODERNISM AND CHRISTIANITY

Shane Weller
MODERNISM AND NIHILISM

Veronica West-Harling
MODERNISM AND THE QUEST

Modernism and...
Series Standing Order ISBN 978-0-230-20332-7 (Hardback) 978-0-230-20333-4 (Paperback)
(outside North America only)

You can receive future titles in this series as they are published by placing a standing order. Please contact your bookseller or, in case of difficulty, write to us at the address below with your name and address, the title of the series and the ISBN quoted above.

Customer Services Department, Macmillan Distribution Ltd, Houndmills, Basingstoke, Hampshire RG21 6XS, England

MODERNISM AND EUGENICS

Marius Turda

*Deputy Director, Centre for Health,
Medicine and Society,
Oxford Brookes University, UK*

First published 2010 by
PALGRAVE MACMILLAN

Palgrave Macmillan in the UK is an imprint of Macmillan Publishers Limited,
registered in England, company number 785998, of Houndmills, Basingstoke,
Hampshire RG21 6XS.

Palgrave Macmillan in the US is a division of St Martin's Press LLC,
175 Fifth Avenue, New York, NY 10010.

Palgrave Macmillan is the global academic imprint of the above companies
and has companies and representatives throughout the world.

Palgrave® and Macmillan® are registered trademarks in the United States,
the United Kingdom, Europe and other countries.

ISBN: 978–0–230–23082–8 hardback
ISBN: 978–0–230–23083–5 paperback

This book is printed on paper suitable for recycling and made from fully
managed and sustained forest sources. Logging, pulping and manufacturing
processes are expected to conform to the environmental regulations of the
country of origin.

A catalogue record for this book is available from the British Library.

Library of Congress Cataloging-in-Publication Data

Turda, Marius.
 Modernism and eugenics / Marius Turda
  p.cm.
 ISBN 978–0–230–23083–5 (pbk.) 978–0–230–23082–8 (cloth)
 1. Eugenics—Europe—History. 2. Modernism (Aesthetics)—
Europe—History. I. Title.

HQ755.5.E85T87 2010
363.9'209409041—dc22   2010002715

10 9 8 7 6 5 4 3 2 1
19 18 17 16 15 14 13 12 11 10

Printed and bound in Great Britain by
CPI Antony Rowe, Chippenham and Eastbourne

To Aliki

CONTENTS

Series Preface   ix

Acknowledgements   xv

Introduction: Context and Methodology   1

1 **The Pathos of Science, 1870–1914**   13
   The Allure of Scientism   13
   The Eugenic Ideal   19
   The Menace of Degeneration   24
   Internationalising Eugenics   31

2 **War: The World's Only Hygiene, 1914–1918**   40
   War as an Educator   42
   Nations Racially Damaged   46
   The Eugenic Crusade: Quantity or Quality?   52
   Political Biology   56
   The Dawn of a New Era   62

3 **Eugenic Technologies of National
   Improvement, 1918–1933**   64
   Eugenic Stigma   65
   Post-War Reconstruction   69
   The Nationalisation of Eugenics   72
   Unworthy Life   79
   Heavenly Foundations, Rational Planning   83

4 **Eugenics and Biopolitics, 1933–1940**   92
   Practical Applications of Eugenics   93
   Eugenic Entopia   100
   Controlling the Ethnic Minorities   107
   The Biopolitical State   110

**Conclusion: Towards an Epistemology of
Eugenic Knowledge**   118

Notes   127

Select Bibliography   153

Index   181

SERIES PREFACE

As the title 'Modernism and ...' implies, this series has been conceived in an open-ended, closure-defying spirit, more akin to the soul of jazz than to the rigour of a classical score. Each volume provides an experimental space allowing both seasoned professionals and aspiring academics to investigate familiar areas of modern social, scientific or political history from the defamiliarizing vantage point afforded by a term not routinely associated with it: 'modernism'. Yet this is no contrived make-over of a clichéd concept for the purposes of scholastic bravado. Nor is it a gratuitous theoretical exercise in expanding the remit of an 'ism' already notorious for its polyvalence – not to say its sheer nebulousness – in a transgressional fling of postmodern *jouissance*.

Instead this series is based on the *empirically*-oriented hope that a deliberate enlargement of the semantic field of 'modernism' to embrace a whole range of phenomena apparently unrelated to the radical innovation in the arts it normally connotes will do more than contribute to scholarly understanding of those topics. Cumulatively, the volumes that appear are meant to contribute to a perceptible paradigm shift slowly becoming evident in the way modern history is approached. It is one which, while indebted to 'the cultural turn', is if anything 'post-post-modern', for it attempts to use transdisciplinary perspectives and the conscious clustering of concepts often viewed as unconnected – or even antagonistic to each other – to consolidate and deepen the reality principle on which historiography is based; not fleeing it, but moving closer to the experience of history of its actors. Only those with a stunted, myopic (and actually *unhistorical*) view of what constitutes historical 'fact' and 'causation' will be predisposed to dismiss the 'Modernism and...' project as mere 'culturalism', a term which due to unexamined prejudices and sometimes sheer ignorance has, particularly in the vocabulary of more than one eminent 'archival' historian, acquired a reductionist, pejorative meaning.

Yet even open-minded readers may find the title of this book disconcerting. Like all the volumes in the series, it may seem to conjoin two phenomena that do not 'belong'. However, any 'shock of the new' induced by the widened usage of modernism to embrace non-aesthetic phenomena that makes this juxtaposition possible should be mitigated by realizing that in fact it is neither new or shocking. The conceptual ground for a work such as *Modernism and Eugenics* has been prepared for by such seminal texts as Marshall Berman's *All that is Solid Melts into Thin Air. The Experience of Modernity* (1982), Modris Eksteins' *Rites of Spring* (1989), Peter Osborne's *The Politics of Time. Modernity and the Avant-garde* (1995), Emilio Gentile's *The Struggle for Modernity* (2003), and more recently Mark Antliff's *Avant-Garde Fascism. The Mobilization of Myth, Art and Culture in France, 1909–1939* (2007). In each case modernism is revealed as the long-lost sibling (twin or maybe even father) of historical phenomena rarely mentioned in the same breath.

Yet the real pioneers of such a 'maximalist' interpretation of modernism were none other than some of the major modernists. For them the art and thought that subsequently earned them this title was a creative force – passion even – of revelatory power which, in a crisis-ridden West where *anomie* was reaching pandemic proportions, was capable of regenerating not just 'cultural production', but 'socio-political production', and for some even society *tout court*. Figures such as Friedrich Nietzsche, Richard Wagner, Wassily Kandinsky, Walter Gropius, Pablo Picasso or Virginia Woolf never accepted that the art and thought of 'high culture' were to be treated as self-contained spheres of activity peripheral to – and cut off from – the mainstreams of contemporary social and political events. Instead they assumed them to be laboratories of visionary thought vital to the spiritual salvation of a world being systematically drained of higher meaning and ultimate purpose by the dominant, 'nomocidal' forces of modernity. If we accept Max Weber's thesis of the gradual *Entzauberung*, or 'disenchantment' of the world through rationalism, such creative individuals can be seen as setting themselves the task – each in his or her own idiosyncratic way – of *re-enchanting* and re-sacralizing the world. Such modernists consciously sought to restore a sense of higher purpose, transcendence, and *Zauber* to a spiritually starved modern humanity condemned by 'progress' to live in a permanent state of existential exile, of *liminoid transition*, now that the forces of the divine seemed to have withdrawn in what

Martin Heidegger's muse, the poet Friedrich Hölderlin, called 'The Flight of the Gods'. If the hero of modern popular nationalism is the Unknown Warrior, perhaps the patron saint of modernism itself is *Deus Absconditus*.

Approached from this oblique angle modernism is thus a revolutionary force, but is so in a sense only distantly related to the one made familiar by standard accounts of the (political or social) revolutions on which modern historians cut their teeth. It is a 'hidden' revolution of the sort referred to by the 'arch'-aesthetic modernist Vincent van Gogh musing to his brother Theo in his letter of 24 September 1888 about the sorry plight of the world. In one passage he waxes ecstatic about the impression made on him by the work of another spiritual seeker disturbed by the impact of 'modern progress', Leo Tolstoy:

> It seems that in the book, *My Religion*, Tolstoy implies that whatever happens in a violent revolution, there will also be an inner and hidden revolution in the people, out of which a new religion will be born, or rather, something completely new which will be nameless, but which will have the same effect of consoling, of making life possible, as the Christian religion used to.
>
> The book must be a very interesting one, it seems to me. In the end, we shall have had enough of cynicism, scepticism and humbug, and will want to live – more musically. How will this come about, and what will we discover? It would be nice to be able to prophesy, but it is even better to be forewarned, instead of seeing absolutely nothing in the future other than the disasters that are bound to strike the modern world and civilization like so many thunderbolts, through revolution, or war, or the bankruptcy of worm-eaten states.

In the series 'Modernism and ...' the key term has been experimentally expanded and 'heuristically modified' to embrace any movement for change which set out to give a name and a public identity to the 'nameless' and 'hidden' revolutionary principle that van Gogh saw as necessary to counteract the rise of nihilism. He was attracted to Tolstoy's vision because it seemed to offer a remedy to the impotence of Christianity and the insidious spread of a literally soul-destroying cynicism, which if unchecked would ultimately lead to the collapse of civilization. Modernism thus applies in this series to all concerted attempts in any sphere of activity to enable life to be lived more 'musically', to resurrect the sense of transcendent communal and individual purpose being palpably eroded by the chaotic unfolding of events in the modern world even if the end result would be 'just' to make society physically and mentally healthy.

What would have probably appalled van Gogh is that some visionaries no less concerned than him by the growing crisis of the West sought a manna of spiritual nourishment emanating not from heaven, nor even from an earthly beauty still retaining an aura of celestial otherworldliness, but from strictly secular visions of an alternative modernity so radical in its conception that attempts to enact them inevitably led to disasters of their own following the law of unintended consequences. Such solutions were to be realized not by a withdrawal from history into the realm of art (the sphere of 'epiphanic' modernism), but by applying a utopian artistic, mytho-poeic, religious, or technocratic consciousness to the task of har-nessing the dynamic forces of modernity itself in such spheres as the natural sciences and social engineering in order to establish a new social, political and biological order. It is initiatives conceived in this 'programmatic' mode of modernism that the series sets out to explore. Its results are intended to benefit not just a small coterie of like-minded academics, but mainstream teaching and research in modern history, thereby becoming part of the 'common sense' of the discipline even of self-proclaimed 'empiricists'.

Some of the deep-seated psychological, cultural and 'anthro-pological' mechanisms underlying the futural revolts against modernity here termed 'modernism' are explored at length in my *Modernism and Fascism. The Sense of a Beginning under Mussolini and Hitler* (2007). The premise of this book could be taken to be Phillip Johnson's assertion that 'Modernism is typically defined as the condition that begins when people realize God is truly dead, and we are therefore on our own.' It presents the well-springs of mod-ernism in the primordial human need for transcendental meaning in a godless universe, in the impulse to erect a 'sacred canopy' of culture which not only aesthetically veils the infinity of time and space surrounding human existence to make existence feasible, but provides a totalizing world-view within which to locate indi-vidual life narratives, thus imparting it with the illusion of cosmic significance. By eroding or destroying that canopy, modernity cre-ates a protracted spiritual crisis which provokes the proliferation of countervailing impulses to restore a 'higher meaning' to histori-cal time that are collectively termed by the book (ideal-typically) as 'modernism'.

Johnson's statement seems to make a perceptive point by associating modernism not just with art, but with a general 'human condition' consequent on what Nietzsche, the first great modernist philosopher, called 'the Death of God'. Yet in the context of this series his statement requires significant qualification. Modernism is *not* a general historical condition (any more than 'post-modernism' is), but a generalized revolt against even the *intuition* made possible by a secularizing modernization that we are spiritual orphans in a godless and ultimately meaningless universe. Its hallmark is the bid to find a new home, a new community, and a new source of transcendence.

Nor is modernism itself necessarily secular. On the contrary: both the wave of occultism and the Catholic revival of the 1890s and the emergence of radicalized, Manichaean forms of Christianity, Hinduism, Islam and even Buddhism in the 1990s demonstrate that modernist impulses need not take the form of secular utopianism, but may readily assume religious (some would say 'post-secular') forms. In any case, within the cultural force field of modernism even the most secular entities are sacralized to acquire an aura of numinous significance. Ironically, Johnson himself offers a fascinating case study in this fundamental aspect of the modernist rebellion against the empty skies of a disenchanted, anomic world. A retired Berkeley law professor, books like *The Wedge of Truth* made him one of the major protagonists of 'Intelligent Design', a Christian(ized) version of creationism that offers a prophylactic against the allegedly nihilistic implications of Darwinist science.

Naturally no attempt has been made to impose 'reflexive metanarrative' developed in *Modernism and Fascism* on the various authors of this series. Each has been encouraged to tailor the term modernism to fit their own epistemological cloth, as long as they broadly agree in seeing it as the expression of a reaction against modernity not restricted to art and aesthetics, and driven by the aspiration to create a spiritually or physically 'healthier' modernity through a new cultural, political and ultimately biological order. Naturally, the blueprint for the ideal society varies significantly according to each diagnosis of what makes actually existing modernity untenable, 'decadent' or doomed to self-destruction.

The ultimate aim of the series is to help bring about a paradigm shift in the way the term 'modernism' is used, and hence stimulate fertile new areas of research and teaching with an approach which enables methodological empathy and causal analysis to be applied even to events and processes ignored by or resistant to the explanatory powers of conventional historiography. I am delighted that its first volume is Marius Turda's groundbreaking and fascinating exploration of the subtle relationship between modernism and eugenics, an association in which utopian biological planning and dystopian results are inextricably connected through the sometimes fatal logic of modernist revolution.

ROGER GRIFFIN
Oxford
January 2010

ACKNOWLEDGEMENTS

I would like to thank Roger Griffin who has been for many years a wonderfully supportive friend. It was his generous offer to contribute to his nascent collection on modernism that led to the writing of this book.

Cristina Bejan in Washington, DC, Michal Kopeček in Prague, Margit Berner and Daniela Sechel in Vienna and Budapest, Francesco Cassata in Turin and Paula Luckett in Oxford were kind enough to provide me with copies of articles and books from libraries I could not re-visit whilst working on this book.

My research was supported by the Wellcome Trust in London and Oxford Brookes University. Carol Beadle, Steve King and Paul Weindling, in particular, were very encouraging. The School of Arts and Humanities at Oxford Brookes University provided a great collegial atmosphere and a stimulating intellectual environment, one that scholars always dream of but rarely experience.

Amongst those who have read drafts of this book, I am hugely indebted, and not for the first time, to Tudor Georgescu, for his assistance and, most importantly, his invaluable critical eye. He is a dedicated friend and colleague. Dan Stone too provided useful suggestions and helped me refine and strengthen some of the crucial arguments discussed in the book. Volker Roelcke also provided valuable suggestions on the book's final conceptualisation. None of them are however to be held responsible for my errors and shortcomings.

Finally, I want to thank my wife, Aliki, for her unfailing moral support. This book is dedicated to her.

INTRODUCTION: CONTEXT AND METHODOLOGY

The history of eugenics, long viewed through the prism of Nazi racial hygiene, has finally reached the level of conceptual maturity necessary for a comparative and multidisciplinary examination. Over the past decades, eugenics was seen as a biological theory of human improvement grounded almost exclusively in ideas of race and class. But eugenics was equally a social and cultural philosophy of identity predicated upon modern concepts of purification and rejuvenation of both the human body and the larger national community. It is only recently that scholars have begun to approach eugenics as a cluster of diverse biological, cultural and religious ideas and practices that interacted with a variety of social, cultural, political and national contexts.[1]

Apart from a continuous focus on the British, American and German variants of a wider European eugenic discourse, current interests have expanded both geographically by covering countries as diverse as Brazil, Russia, Romania, Turkey and China, and thematically by unravelling the important connections between eugenics and population policies, as well as its relationship with a number of political ideologies, including nationalism, liberalism, social democracy, anarchism, communism and fascism.[2] Given this innovative conceptual framework, it is surprising to note that hardly any comparative research has been conducted into the relationship between modernism and eugenics. Existing works largely concentrate on how eugenic ideas permeated modernist literary culture (and particularly so in Britain[3]), leaving an important aspect of this relationship unexplored: the modernist engagement with eugenic theories of human improvement and eugenic visions of national perfection. It is this insufficiently explored aspect of the history of European eugenics that I am pursuing in this book.

The current affluence of scholarship on modernity, and the interdisciplinary convergence it generated, has not only furthered an increased awareness of diverse traditions of eugenic thought, both defunct and active, but also prompted scholars to carefully examine established currents of thought shaping these traditions. Drawing inspiration from this scholarship this book argues that eugenics should be understood not only as a scientistic narrative of biological, social and cultural renewal, but also as the emblematic expression of programmatic modernism. That is, the form of modernism that "encourages the artist/intellectual to collaborate proactively with collective movements for radical change and projects for the transformation of social realities and political systems."[4] Roger Griffin applied this broad conceptual model to suggest an eclectic interpretation of fascism, seeing it as the main consequence of the European society's yearning for a new beginning. Racial improvement was an intrinsic component of this generalised longing for new foundations, with eugenics as the most sophisticated attempt to improve individuals and societies through biological engineering. It is in this strict, minimal sense, I argue, that eugenics intersected political ideologies like communism and fascism. To put it differently, and by alluding to one of Walter Benjamin's classical aphorisms, whereas communists politicised art and fascists aestheticised politics, eugenicists biologised identity.

To substantiate the claim that eugenics is both a cluster of scientistic narratives and an expression of modernity, this book looks at debates and speculations on the nature of the relationship between modernist thought and eugenics in various European countries between 1870 and 1940. During this period, Europe went through profound territorial, social and national transformations, and experienced a wide range of political systems in rapid succession: from imperial to democratic, communist, authoritarian, and fascist. As a corollary of this seismic transformation of the European political landscape, eugenics became part of a larger biopolitical agenda that included social and racial hygiene, public health and family planning as well as racial research into social and ethnic minorities. In this context, eugenics was as diverse ideologically as it was spread geographically, both advocated and adhered to by professional and political elites across Europe irrespective of their political and intellectual affiliation.

In order to capture the versatile nature of eugenics, this study will draw upon two methodological clusters. The first of these clusters highlights the broadness of eugenic thought as, like most modern political and social ideologies, eugenics comprised a wide range of views. Rather than looking at eugenics as a rigid discursive structure centred on race, gender or class, one must approach it instead as a polysemic system of thought.[5] As Lene Koch noted, eugenics was "not a fixed, well-defined ontological entity with one definite purpose," but "a social practice that constituted a complex array of goals and viewpoints that cannot be condensed into one single meaning. As in the case with the concept of 'science', one single definition would necessarily be arbitrary and of no use in a historic study."[6]

The second methodological cluster subscribes to the practices of conceptual intellectual history (*Begriffsgeschichte*). As Melvin Richter explained, "*Begriffsgeschichte* puts on the agenda of practicing historians questions about the significance of change and continuity, of contestation and consensus about meaning in political and social languages."[7] It is this understanding of eugenics as a contested historical concept that underscores this study's structural morphology and points equally to the interrelationships between eugenic theory and practice. The history of eugenics cannot, therefore, be written without paying attention to the political discourses and the specific national cultures in which eugenic ideas were formulated and defined. Robert Nye intuited this when presenting the history of eugenics as "cultural history, where biomedical ideas of various provenance are mediated by the social influences of institutions, political power and public opinion, the peculiarities of individual personalities, and by the inexorable logic of geopolitics."[8] By investigating the main arguments of such diverse authors as those presented in this study, it becomes possible to focus on the specific ways in which broader cultural and political trends, such as the avant-garde, racism, nationalism, fascism and Nazism, operated through the medium of science and politics, gradually affecting the way eugenic thought was defined in modern European history until 1940.

In terms of its geographical coverage, this book looks at particular eugenic trajectories in countries as diverse as Britain, Norway, France, Germany, Czechoslovakia, Hungary, Romania, Italy, Spain and Greece. These are countries with different historical, political and cultural traditions as well as different dominant religions. All are predominantly Christian, but some are Catholic, others are Protestant, some are both, while a few are Orthodox. If one adds Turkey to this picture, and the Muslim minority in interwar Yugoslavia, a heterogeneous ensemble of countries and nations emerges, one where each advocates its own distinctiveness, yet sub-scribes to the regenerative force of eugenic modernism. A number of questions arise in this context: how was the nation defined in these countries? Was it an "imagined community", unifying individuals and groups around shared linguistic, political, cultural and religious characteristics; or rather – as the eugenicists argued – a community of blood and genes, determined by heredity? Is there, furthermore, something that unites these countries, in terms of their eugenic quest for national regeneration?

To be sure, the local socio-economic specificities were also crucial in determining how eugenic principles were perceived and translated into practice, as countries like Britain and Germany, for example, displayed a far higher degree of industrialisation and urbanisation than the largely rural and agrarian case studies offered by Romania and Greece. Yet the source of national rejuvenation was similar in all cases, namely the natural environment and peasantry (in its various permutations), both of which were deemed healthy and uncontami-nated by the ills of modernity. The pursuit of a healthy national body was consequently central to eugenic discourses across Europe, and throughout the period under investigation here, irrespective of how different eugenic movements appeared to be in practice. An analysis of these highly diverse eugenic discourses offers new means through which to explore the transformation of eugenics, from the product of a particular national environment into a trans-national modern-ist philosophy.

Moreover, while the Western European histories of eugen-ics are well researched, little is known about eugenics in Central and Southeastern Europe. Until recently, these regions have been neglected by scholarship despite having much to offer in terms of advancing our understanding of the connections between modern-ism and eugenics in various political and cultural contexts.[9] There is substantial literature on the relationships between British and American eugenicists; the connections between German racial hygiene and its Scandinavian counterparts; the impact American eugenics had on German racial hygiene, the relationship between German and Soviet eugenicists; and the attempts by Italian, French and Spanish eugenicists to build a strong counter-narrative to what they perceived as increasingly racist "Nordic" eugenics. Yet little is know about the role Central and Southeastern European eugeni-cists played within this international transfer of knowledge and exchange of practices and ideas. Scholars working on French and Italian eugenics, for example, acknowledge the Romanian contribution to the "Latin" version of international eugenics. But what was this contribution? Characteristically, even the existing scholarship on Romanian eugenics has been unable to explain its relevance to the larger, international, context. This is another aspect of the history of European eugenics that this book aims to illuminate.

An asymmetric comparison is, ultimately, needed in order to evaluate different national contexts that nevertheless shared similar eugenic ideas and practices. The main sources for this aggregated comparative overview are primarily specialised journals, proceedings of national and international conferences on eugenics, scholarly works and books geared towards popularising eugenics amongst the general public. This eclecticism is meant to reinforce the idea that eugenics served not only to generate medicalised metaphors of the social and national body, but to augment those technologies of hygiene and health without which modern societies were allegedly destined to immerse in barbarity and backwardness. To look at these technologies closely is to confront what Alison Bashford describes as the "political and cultural imagining of bodies and nations."[10] This imagining is central to modernism and eugenics alike, as both incorporate the new ideal of humanity into their revolutionary programme of social and national purification.

This investigation's ultimate purpose is not, therefore, to offer an exhaustive history of European eugenics, but to examine how eugenic movements influenced visions of national regeneration as well as biopolitical ideologies and models of racial and social engineering. To achieve this goal, the book is pursuing two conceptual aims: first, it examines the methodologies through which the individual body was re-defined eugenically by a diverse range of scientists, intellectuals and politicians; and, second, it illuminates how the collective body – the nation – was represented by the eugenic and biopolitical discourses that strove to battle a perceived process of cultural decay and biological degeneration.

By the end of the nineteenth century, nations were increasingly being portrayed as living organisms, functioning according to biological laws, and embodying great genetic qualities symbolising innate racial virtues transmitted from generation to generation.[11] After 1900, especially, this shifting relationship between the individual and the racial community to which he or she belonged contributed significantly to the emergence of a eugenic ontology of the nation. As a result, the individual body became a synecdoche for the collective national body depicted as an organism susceptible to the biological debilities that attend birth, growth, aging and dying. As a quintessential modernist response to threats of cultural and biological collapse, eugenics located the individual and national body within a specifically scientific discourse, one whose legitimacy stemmed from its preoccupation with improving the racial quality of the population and protecting its health.

With the emergence of the science of genetics around the turn of the twentieth century, a new sense of identity has been constructed around biology. In his inimitable style, Michel Foucault termed this process of biological appropriation the "political technology of the body," indicating the crucial importance modernity bestowed on the relationship between political power and scientific knowledge.[12] Within this newly formed biopolitical epistemology, eugenics emerged as one of the most convincing answers to a series of social, economic and political crises characterising European modernity since the late nineteenth century. Moreover, the eugenic discourse – like modernism itself – was deeply embedded in what Reinhardt Koselleck has termed the "Neuzeit," that is the beginning of a new historical time in European culture. By "being able to begin history anew,"[13] when troubled by the prospect of racial dissolution and national defeat, the individual and the community found in eugenics a persuasive strategy of how to protect the past from a dissatisfying present, and how to guide it into a redeeming future.

Yet, this development of eugenics impacted not only the representation of the body in the modernist imagination, but also transformed the nation into an object of scientific regulation and expertise. This biologisation of national belonging was to have a tremendous impact on the evolution of eugenics, especially after the First World War. The nation was progressively portrayed as a biological entity whose natality, longevity, morbidity and mortality needed to be supervised. At a time when theories of biological destiny enthralled political elites and intellectuals alike, eugenicists offered the possibility of national regeneration, combining scientific dogmas with racial categories, thus illustrating what Roger Griffin described as an "alternative modernity," the essential condition to the nation's rebirth or palingenesis.[14]

Eugenics embodied, in contrast to other modernist strategies devised to salvage the community from cultural decadence and biological degeneration, a fusion between the biological finitude of the individual and the eternal existence of the nation. By demanding that the state should aim for the purification of its racial body through eugenic means, eugenicists thought of themselves as working towards a noble goal: the creation of a strong and healthy nation. Therefore, they were not simply preoccupied with rescuing the individual from the anomie of modernity; they geared their efforts towards saving the nation and assuring it a luminous racial destiny.

It should be noted, moreover, that eugenics emerged not only as a biological critique of a degenerative modernity, or as a process of political, legal and institutional control over the population contained within a delimited territorial space, but also as a defensive strategy for a particular race. Prompted by the need to generate a powerful sense of cohesion and shared identity in the wake of perceivably profound structural social changes, modernist eugenicists appealed to racial imagery in order to justify their biologisation of national belonging. On the one hand, any race's identity was determined by biological, social and cultural boundaries that separated those who belonged to the community from foreigners and outsiders who remained aliens or potential enemies. On the other hand, however, eugenics also created a system of internal sanitisation according to which those deemed "unhealthy," "diseased" or "antisocial" were separated from the "healthy" majority, segregated, and, in some cases, subjected to radical measures such as sterilisation and euthanasia. As the British sexologist and eugenicist Havelock Ellis powerfully expressed it:

> If in our efforts to better social conditions and to raise the level of the race we seek to cultivate the sense of order, to encourage sympathy and foresight, to pull up racial weeds by the roots, it is not that we may kill freedom and joy, but rather that we may introduce the conditions for securing and increasing freedom and joy. In these matters, indeed, the gardener in his garden is our symbol and our guide.[15]

So whilst different models of practical eugenics emerged in Europe between 1870 and 1940, all were based on three principles: first, the crucial role of heredity in determining the individual's physical condition; second, the link between biology, medicine and the health of the nation; and, third, the politicisation of science. These three

principles would subsequently be tested during the seismic upheav-
als of the First World War, prefiguring the emergence of new forms
of the biologisation of national belonging during the interwar period
and the Second World War.

The reconfiguration of the traditional private sphere along with
individual and religious rights was an important consequence of
this transformation. Essentially, the boundary between private and
public spheres was blurred by the idea of public responsibility for
the nation, which came to dominate both. Biological refashioning
of individual identity was another consequence. David Horn has
described the individuals resulting from these eugenic construc-
tions of the nation as "social bodies, located neither 'in nature' nor
in the private sphere, but in that modern domain of knowledge and
intervention carved out by statistics, sociology, social hygiene, and
social work."[16] Ultimately, eugenics operated through investigations
of biological and social processes regulating the triadic relationship between the individual, the nation and the state. Therefore, one
should not treat eugenics as an extraordinary episode distinct from
the progressive development of the natural and medical sciences, as
a deviation from the norm and a distorted version of crude social
Darwinism that found its culmination in fascism and Nazi policies
of genocide, but as an integral aspect of European modernity, one in
which the state and the individual embarked on an unprecedented
quest to renew an idealised national community.[17]

*   *   *

This exploration of the impact eugenic discourses had on national
cultures in Europe between 1870 and 1940 is structured thematically.
To briefly justify this decision, the sheer volume of sources pertaining to each individual country makes it simply impossible to include
it all in one volume. I have, therefore, chosen the following themes as
they were particular prevalent and pronounced amongst eugenicists
from different European countries, in addition to their coalescing
into a recognisable narrative of modernism.

Chapter 1 investigates the birth of eugenics in its modern form,
and identifies a number of the most important eugenic themes at the
end of the nineteenth and early twentieth century. Much has been
written on Francis Galton's contribution to the emergence of eugenics. Following a series of publications during the 1870s and 1880s,

Galton had embarked on the formulation of a new scientific philosophy based on the theory of natural selection and inheritance, which he termed eugenics, following an entrenched Victorian penchant for ancient Greek culture. He had hoped that the old order, sanctified by religion and outdated scientific dogmas would finally be replaced. Galton had, in fact, hoped that with secularisation and the growing acceptance of theories on evolution and heredity, eugenics would become the scientific doxology of the future.

But how was Galton's eugenics interpreted and adopted in Britain and abroad? How did the Norwegian, Italian, Czech or Hungarian supporters of eugenics react, for instance, to competing models of eugenic thought, like the German ideas of racial hygiene and racial improvement and the French theories of puericulture? This multiple terminology suggests that doubts may have existed amongst supporters of eugenics across Europe as to whether the concept of eugenics was sufficiently illuminating to encompass competing perspectives on social and racial improvement.

On the practical level, this process of conceptual negotiation, appropriation and negation prompted two broader questions about the role of eugenics in shaping views on the biological development of society: How could scientistic paradigms, like eugenics, be made compatible with the social and national particularities? And, secondly, could biologists and physicians be trusted as a source of scientific enlightenment amidst profound social and national transformation? By simultaneously raising the question of scientific legitimacy and demands for practical action in the name of science, the supporters of eugenics challenged the cultural and political establishment to react more resolutely to the social problems that, they argued, had been troubling their respective societies since the late nineteenth-century.

Chapter 2 sets out to explain the connection between eugenics and modernist diagnosis of destruction and resurrection between 1914 and 1918. An analysis of the relationship between eugenics and war offers us a far more insightful avenue through which to understand how scientific ideas on health and hygiene were couched in nationalist and racial idioms, and the means by which these idioms became embedded in social-political agendas. If prior to the outbreak of the First World War eugenicists from various countries had been united in a form of internationalism culminating in the First International Congress on Eugenics convened in 1912 in London, many of them engaged in national politics during the war, devising eugenic methodologies to serve the ideological imperatives of their own countries rather than the proclaimed universalism of the pre-war years.

After 1918, issues of public health, hygiene, eugenics, demography and population control became paramount in the national politics of European states. According to the post-war modernist mythopoeia, the national community was to be fortified not merely under the banner of a cultural and political ideology, but through a new synthesis of biological and eugenic morality. The emergence of national eugenics in a number of European countries was, in short, the answer to a society searching for new foundations upon which to base individual and collective improvement.

Chapter 3, therefore, looks at the relationship between modernism and the practical application of eugenics to society. Since the middle of the nineteenth century, prominent physicians, biologists, social scientists, religious and political leaders have all consistently expressed eugenic views and projects on how to protect the nation and the race from an alleged biological degeneration; on how to improve the health of the population; and on how to increase the number of healthy families. Across Europe, eugenicists were interested in both positive and negative eugenics: where the former concentrated on policies encouraging those deemed healthy to have more children, the latter aimed to discourage and prevent the "unfit" from reproducing.

Between the 1930s and 1940s, laws authorising negative eugenic practices like marriage regulations, sexual segregation, and sterilisation were introduced in most European countries. The debate about eugenics also widened. Hitherto restricted to medical specialists, eugenics now increasingly attracted other categories of professionals, especially religious leaders, legal experts, sociologists and statisticians and it was as passionately debated in Britain and Germany as in Romania, Hungary and Greece, amongst others. In many respects, support for practical eugenics was not simply a symptom of racism and anti-Semitism, but the expression of the desire to protect the national body through controlling its biological foundations. The ultimate eugenic goal was to create a new nation through an all-encompassing act of both individual and collective regeneration. Returning to Foucault's "political technologies of the body," eugenics made it possible for a great number of politicians and intellectuals across Europe to speak of harnessing the national body; cultivating and weeding out extraordinary individuals; and purifying the race. The eugenic vocabulary thus overlapped with an adjacent set of fears over racial and national decline, amplifying the vision of each individual country perceived as "under attack" by internal and external enemies.

Chapter 4, appropriately, engages with the relationship between eugenics and biopolitical representations of the nation. The conclusion of the war in 1918, and the collapse of empires and the creation of new nation-states that followed only increased the potential fragmentation of European culture and politics. Eugenics was channelled into a totalising and centralised heterotopia: the creation of an ideal national community and state. Whether to substantiate British claims of racial prowess, to restore Germany's former historical prestige, to give substance to the political project of Greater Romania, or to assure the survival of the Hungarian nation, eugenics became a cluster of ideas possessing multiple, fragmented and even incompatible meanings. Racial nationalism, in its various forms, built the bridge linking notions of biopolitical welfare with the eugenic imperative.

According to a particularly prevalent eugenic diagnosis, the nation's biological body was constantly threatened by the fear of contamination resulting from the dangerous mixing of different, often opposing, ethnic and biological categories. Translated unto the geopolitical stage of the interwar period, this perception created a certain biopolitical dynamic where eugenic anxieties about the future of the nation and race were intimately associated with fears of absorption by neighbouring nation-states, and by the potentially detrimental impact of growing enclaves of domestic ethnic minorities. During the 1920s and 1930s, eugenics aimed to sanitise this conflicting nationalised space, proposing a new vision of the national community, one biologically purged of all symptoms of Otherness. It is important to understand how this vision of national perfection was constantly recycled and actively reinvented, reaching its climax during the Second World War.

As modern states became increasingly obsessed with their historical mission, namely to create a rejuvenated nation which was culturally, spiritually and racially homogeneous, they also resorted to coercive mechanisms – such as stigmatisation, discrimination, segregation, sterilisation, and ethnic cleansing – in order to protect its chosen members and eliminate those deemed socially, culturally and ethnically damaging to the body of the nation. Eugenics, with its objectifying, materialising, clinical gaze, contributed to this vision of human perfection in which individuals and ethnic groups deemed dangerous to the nation were relegated to institutions and marginal social spaces reserved for the elderly, the disabled and the sick. Whether eugenicists thought in terms of purifying the nation of "defective genes," or protecting it from mixing with "racially inferior" elements, there was widespread agreement that scientific thinking was indispensable to legitimising and rationalising the social engineering and the biopolitical transformation of the nation-state. It is this troubled epistemology of modernity that needs a richer historical understanding if one is to make sense of the relationship between modern political ideologies, like fascism and communism, and eugenics.

Finally, within current debates on "liberal" and "neo-eugenics," this book has the explicit purpose of charting unstable territories in the history of science. Once the synergy between modernism and eugenics is properly conceptualised, historically and scientifically, a more nuanced interpretation of the general relationship between science and politics will emerge. Not surprisingly, recent debates on eugenics' relevance to the post-modern world have not merely served academic purposes. Aware of the general sensitivity surrounding this subject, scholars have attempted to disassociate current debates on genetic engineering from pre-1940 eugenics.[18] In addition to contributing to new cultural and social histories of modernity, this book, consequently, reflects these current, popular and scientific, discourses on the role of contemporary science in shaping individual and collective identities by suggesting that the political and ideological history of eugenics is fundamental to any informed assessment of modern-day debates on population control, fertility, sexual reproduction and ideas of human perfectibility.

The history of modernism and eugenics powerfully illustrates how seemingly universal scientific ideas about human improvement were transformed through a convoluted process of negotiation, refutation and appropriation. As we continue to investigate the origins of twentieth-century totalitarianisms, the complicated history of eugenics not only serves as an example of the exploitation of science for political reasons, but also indicates that there was something extremely appealing in the eugenic promise to design and control human populations, a prospect which all political and cultural ideologies found irresistible.

1

THE PATHOS OF SCIENCE, 1870–1914

The Allure of Scientism

At a time when biological theories of human nature were at their zenith and endorsed by some of the most powerful political regimes in Europe, the Spanish philosopher José Ortega y Gasset unhesitatingly declared that "[m]an has no nature. What he has is history."[1] This was a powerful critique of the biologisation of national identity his contemporaries seemed to accept so eagerly. Indeed, by the late 1930s, most cultural commentators and philosophers – not to mention scientists – were building a consensus that human destiny was determined by evolution and heredity. The French biologist Alexis Carrel helped to popularise this conception of life by describing the scientific achievements of the twentieth century in terms of the ultimate transformation of man: "Science, which has transformed the material world, gives man the power of transforming himself. It has unveiled some of the secret mechanisms of his life. It has shown him how to alter their motion, how to mould his body and his soul on patterns born of his wishes. For the first time in history, humanity, helped by science, has become master of its destiny."[2] This widespread biological understanding of culture was an essential prerequisite for the emergence of a modernist version of eugenics, set to embark upon ambitious policies of human improvement. However, in order to understand the conceptual transformation within this scientific knowledge about human nature, we must first attempt to grasp how eugenics was formulated and disseminated during the end of the nineteenth and at the beginning of the twentieth century.

The new scientific ethos that emerged towards the end of the nineteenth century, and which gained considerable influence during the interwar period, was appropriately termed scientism. In 1942, the social scientist Friedrich A. Hayek, for instance, considered it to be "the slavish imitation of the method and language of Science;"[3] whilst the political theorist Eric Voegelin, a few years later, treated it more appropriately as "a decisive ingredient in modern intellectual movements like positivism and neopositivism, and, in particular, like communism and national socialism."[4] It is in this latter capacity that the cultural philosopher Tzvetan Todorov used scientism, whilst adding that science "or what is perceived as such, ceases to be a simple knowledge of the existing world and becomes a generator of values, similar to religion; it can therefore direct political and moral action."[5] According to this line of thought, scientism was a surrogate for religion in an age of increased atomisation of the social and political life and emerging totalitarianisms. A more nuanced definition of scientism is offered by the historian of science Richard Olson, who discusses it in terms of the "transfer of ideas, practices, attitudes, and methodologies" from the domain of natural sciences "into the study of humans and their social institutions."[6]

The common ground between these convergent arguments is an emphasis on the fusion between scientific language and forms of religious and political rituals. These authors have been able to show how scientific discourses once restricted to a selected number of practitioners and cultural elites had tunnelled galleries into the religious edifice of European culture since the sixteenth century, contributing significantly to the emergence of a hybrid form of knowledge: one which demanded the same loyalty and sacrifice from its followers as religion, but its promised immortality came not from the heavenly church but from the temple of science. "It is we who are, more immediately, the creators of men," Havelock Ellis wrote in his 1911 *The Problem of Race-Regeneration*; and in identifying the scientists as demiurgic creators of man, Ellis emphasised the wondrous connections between religion, history, social and biological engineering in the eugenic aesthetics of the twentieth century: "We generate the race; we alone can regenerate the race."[7] Science, the English zoologist Edwin Ray Lankester added, had become the most expressive medium through which to negotiate the boundaries between a degenerated present and a healthy future. Most importantly, science offered the long-needed assurance for the "protection of our race," affording solutions to its "relapse and degeneration."[8] As incarnations of a new order, scientists were entrusted with safeguarding the community's racial body.

Nineteenth- and twentieth-century nationalists added a new avenue along which to integrate scientism into European culture, namely the nation's redemption and rejuvenation. If the messianic vision of a technologically controlled society derived from scientism, the idea of a purified national body it served belonged to nationalist eugenics. These postulates are now central to the study of fascism and communism,[9] but they still have to penetrate deeper into the broader historiography of modernism and eugenics. Peter Bowler's description of eugenics as "the original political expression of the ideology of genetic determinism"[10] and Michael Burleigh's assumption that eugenics "had evolved from primitive utopianism into a secular religion with scientific pretensions"[11] are indications that such a reading of eugenics is not only possible, but necessary.

In this book I intend to take these interpretations further, arguing that scientism resituates the contribution of eugenics to the history of both European modernism and the revolutionary projects proposed by fascism and Nazism in a number of ways. While we may concur with Todorov that scientistic discourses were incited and activated as instruments of power by totalitarian regimes in the twentieth century we may add that, within this schema, it was the eugenic vision of a healthy nation that entailed the *actual* transformation of both bodies and minds. Thus, while eugenics was indeed designed to illustrate contemporary political ideologies, and did so by remodelling the individual and the society, it also illustrated something else, namely the implication of transcendence, accompanying the knowledge of how man's biological transformation could, in fact, be achieved.

The language of eugenics was, from the outset, situated within the climate of the late-nineteenth-century interaction between religion and science. In terms of science, this period was rather eclectic with various diverse and intersecting hereditarian trends – of which Lamarckism, Darwinism, Weismannism, and Mendelism were the most prominent. One of the salient points of disagreement amongst these theories lay with the ongoing nature vs. nurture debate, namely the preponderance bestowed upon the environment in shaping the hereditary structure of the individual.[12] Biologists in the nineteenth century, following the theory of acquired characteristics developed by Jean Baptiste Lamarck, had accorded greater importance to natural and social environments than to heredity. Even the architect of the theory of natural selection Charles Darwin accepted the role played by environmental factors in the work of the natural selection as paramount.

Another crucial moment was the rediscovery of Gregor Mendel's studies of inheritance in 1900. The largely ignored model of heredity Mendel had published as early as 1865 – later known as the system of Mendelian inheritance – challenged the theory of blending inheritance (namely that an offspring's makeup constituted a blend of his her parents' traits), one which Darwin and Galton, in fact, defended. Mendel's research on the nature of inheritance proved, however, that an offspring's genetic structure was not simply a synthesis of that of the parents, but that particular genetic traits could either be dominant or recessive, and that these were passed on in accordance with mathematical laws.

Some early twentieth-century eugenicists such as Francis Galton and his disciple Karl Pearson were uncomfortable with the Mendelian laws of inheritance, offering a statistical study of the range of hereditary variation across the population instead. Galton's concept of heredity – his "law of ancestral heredity" – rested on the assumption that hereditary characteristics were not only produced by the parents, but transmitted unaltered down the generations. In his 1865 paper on "Heredity Character and Talent" Galton argued that heredity was influenced not by the parents but by the constitution of the ancestral group established over many generations. With respect to talent, his theory asserted that "if talented men were mated with talented women, of the same mental and physical characters as themselves, generation after generation, we might produce a highly-bred human race, with no more tendency to revert to meaner ancestral types than is shown by our long-established breeds of race-horses and fox-hounds."[13]

In his remarks here, as in his insistence elsewhere on the line of an ancestral original racial type, Galton was careful to place limits upon the incidental improvement brought about by better nurture and education, social metamorphoses that he did not deem sufficient to shape "the noble qualities" of future generations. "Can we hand anything down to our children that we have fairly won by our own independent exertions?" Galton asked. "Will our children be born with more virtuous dispositions, if we ourselves have acquired virtuous habits? Or are we no more than passive transmitters of a nature we have received, and which we have no power to modify?" The answer provided encapsulated Galton's renunciation of the theory of acquired characteristics: "If the habits of an individual are transmitted to his descendants, it is, as Darwin says, in a very small degree, and is hardly, if at all, traceable."[14]

For Galton, the biological determinist, the effects of education and environment on the improvement of the race were therefore negligible, while physical and social appearances of the individual were often deceiving. "It is objected," he noted in his 1873 article "Hereditary Improvement" that "any prospect of improving the race of man is absurd and chimerical, and that though enquiries may be pursued for the satisfaction of a curious disposition, they can be of no real importance."[15] Against these views, Galton proposed a programme of racial improvement "entirely based on the assumption that the ordinary doctrines of heredity are, in a broad sense, perfectly true." Elaborating on the necessity to synchronise morality and science, Galton praised not only the importance of heredity in determining the life of individuals, races and nations, but also the role of the eugenicists, who "shall come to think it no hardheartedness to favour the perpetuation of the stronger, wiser, and more moral races."[16] To improve the race should become a systematic, ritualised practice, Galton recommended. The eugenic harmony of race will be restored after a few generations. "The earlier results will be insignificant in number, and disappointing to the sanguine and ignorant," Galton concluded, "who may expect a high race to be evolved out of the present mongrel mass of mankind in a single generation. Of course this is absurd; there will be numerous and most annoying cases of reversion in the first and even in the second generation, but when the third generation of selected men has been reached, the race will begin to bear offspring of distinctly purer blood than the first, and after five or six generations, reversion to an inferior type will be rare."[17]

Crucial to this narrative of racial improvement was, as Angelique Richardson suggested, "genealogy, life history; historical records which might overwrite or underwrite the stories the body could tell. After all, hereditary taint might conceal or misrepresent itself."[18] This singular focus on the history of the individual, on origins and ancestry, was repeated habitually whenever the corporeality of the eugenic subject was questioned. The body, for eugenicists, was thus a heterogeneous synthesis of physicality and history in which the ephemeral biological condition of the present intersected previous trajectories of inheritance. Correlatively, blood came to represent not only the body's most symbolic component, but also the biological closure of eugenics.

Following Karl Landsteiner's discovery of the blood groups (A, B, O) around 1900, the Polish microbiologist Ludwik Hirszfeld confirmed that the percentage of blood groups in a population varied according to their respective racial origins. These authors not only helped sustain the emergence of serology as a discipline preoccupied with deciphering the chemical properties of blood groups for the benefit of improving medical care (such as blood transfusions), and the discovery of new vaccines, but also brought the fascination with blood into the mainstream of the racial imagination. The idea of "biochemical races," as Hirszfeld called them, echoed particularly widely and provided eugenics with a new method for classifying human groups by more accurate, biochemical means rather than the highly contested anthropometric characteristics advocated by anthropology.[19]

Equally important, serology also demonstrated that blood groups were inherited according to Mendelian laws of heredity, thus impregnating the individual with one distinguishing attribute, one which was impervious to internal or external influences. As racial measurements have proven incapable of providing definitive answers to historical questions about racial identity, eugenicists hoped that heredity and serology could offer the scientific certainty needed to legitimise theories of biological uniqueness. The eugenic aesthetics that emerged thus had broader implications for the understanding of programmatic modernist theories of the nation and race when cast in terms of the dichotomy between the perennial nature of blood and the ephemeral, atavistic impact of culture and history.

Considering these discoveries in the natural sciences and biology, it is no surprise that numerous scientists bravely posed as creators of new values and morality. In the place of the spiritual model of identity provided by religion one now found the progressive refashioning of the biological identity provided by science and eugenics. As the secretary of the Zoological Society of London, Peter Chalmers Mitchell declared in 1903, "[i]n every country, the new Order of priests of science, in the vigils of the laboratory, is working for the future of humanity."[20] In this sense, religion and science were not antithetical but complementary activities coupled in a synergetic relationship, one upon which the eugenic ideal of the twentieth century was based.

The Eugenic Ideal

Eugenics was, indeed, an intrinsic part of this scientistic ethos and aspired to offer a doxology of racial development that embraced humanity as a whole, and reconciled theories of particular, national evolution with a commitment to a new future. Intellectual perseverance, in addition to the new scientific ethos, was equally necessary for the transformation of eugenics from a specialised scientific discourse into a topic of public interest. Accordingly, eugenicists viewed themselves not merely as scientists in the narrow sense, but as champions of a new form of intellectual and cultural activity that sought to find a balance between scholarly detachment and political activism.

No one exemplifies this characterisation better than Francis Galton, the "founder of the eugenic faith."[21] Like many Victorian scientists, Galton was interested in a wide range of subjects, from geography to phrenology, and also impressed as an inventor. But the passion for eugenics was central to this polymathic personality. Believing that heredity and not environmental influences could explain the racial, cultural and social differences between populations, Galton applied quantitative measurements and analysis of variants in the physical dimensions of the human body to explain social class differentials in British society.

In 1869, Galton published *Hereditary Genius*, a book that prompted his half-cousin Charles Darwin to exclaim: "I do not think I ever in all my life read anything more interesting and original."[22] Without referring to eugenics directly, the book discussed the origins of natural ability in humans and various possibilities for racial betterment at length. It was, however, in *Inquiries into Human Faculty and Its Development*, a book published in 1883, that Galton engaged with "various topics more or less connected with that of the cultivation of the race, or, as we may call it, with 'eugenic' questions." A more detailed explanation of his definition of eugenics was offered by the note accompanying the text. Eugenics dealt with, "questions bearing on what is termed in Greek, *eugenes*, namely, good in stock, hereditarily endowed with noble qualities."[23] Just as Darwinism may be seen as challenging the hegemonic role of religion and the biological fixity of the human species, eugenics may be seen as supporting the very notion of humanity as defined in terms of a hierarchy of distinct social bodies, some better biologically equipped than others. "We greatly want a brief word to express the science of improving the stock," Galton confessed,

> which is by no means confined to questions of judicious mating, but which especially in the case of man, takes cognisance of all influences that tend in however remote a degree to give to the more suitable races or strains of blood a better chance of prevailing speedily over the less suitable than they otherwise would have had. The word *eugenics* would sufficiently express the idea; it is at least a neater word and a more generalised one than *viriculture*, which I once ventured to use.[24]

To create the appropriate social and national culture for each individual was of particular concern to eugenicists, and Galton proposed to invigorate the national organism and alleviate its existing social pathologies. Eugenics was a new form of biological knowledge which, from its inception, advocated interventions in both public and private spheres. In the name of a healthy national body, eugenicists greatly expanded on biological and social discourses on health. But these discourses differentiated between "worthy" and "unworthy" members of the community. In his seminal 1895 text on racial hygiene, the German eugenicist Alfred Ploetz, thus, argued against pursuing those methods of social selection that encouraged the "weak" to survive and reproduce. "Individual hygiene," Ploetz claimed, was to be subordinated to the "principle of racial hygiene."[25]

The remedies eugenicists proposed to improve the quality and quantity of healthy individuals were linked to the representation of eugenics as both secular religion and science. Galton emphasised both aspects in his celebrated 1904 discussion of eugenics. The time had come, Galton argued, to establish the groundwork for the general reception of eugenics. First, he contended, eugenics "must be made familiar as an academic question." Eugenicists regularly complained that the general public, and even numerous scientists, were yet to be persuaded of the scientific importance of hereditarian theories of human improvement. One strategy to be pursued towards this goal, Galton continued, was to have eugenics "be recognized as a subject whose practical development deserves serious consideration," namely the convergence of the biological with the political into a practical programme of social engineering. Building on these achievements, then, one could hope to introduce eugenics "into the national conscience, like a new religion."[26] Set alongside his preference for elite rule, his instinctive disapproval of the "lower" classes, and his strong commitment to heredity, Galton's eugenic religion seemed as much an expression of traditional class protectionism as of scientism.

Eugenics, consequently, had "indeed strong claims to become an orthodox religious tenet of the future, for eugenics co-operate with the workings of nature by securing that humanity shall be represented by the fittest races. What nature does blindly, slowly, and ruthlessly, man may do providently, quickly, and kindly."[27] In appropriating the authority of religion for eugenics, Galton ventured to ponder what religion might be without the divine design whilst, at the same time, demanding that eugenics be seen to challenge the premises underlying other scientific disciplines dealing with the body of the individual, such as sociology and anthropology. In his 1905 "Studies in Eugenics", Galton reiterated his commitment to a dynamically modernising theory of racial improvement, viewing it within a new biological ontology. Incidentally, it was with this occasion that the now familiar concise definition of eugenics was offered, namely: "Eugenics may be defined as the science which deals with those social agencies that influence, mentally or physically, the racial qualities of future generations."[28]

So, what did Galton's characterisation of eugenics as the "religion of the future" really amounted to? Its central message was that social protection and assistance, including philanthropy and charity, had developed at the expense of the nation's racial qualities, and that this imbalance needed to be remedied, urgently, and especially so with regards to religious and moral education. But, more importantly, this eugenic philosophy was predicated upon a view of individual and social regeneration regulated by science. Eugenics, from this perspective, was not about encouraging the individual's public involvement as it had been in the individualist liberal tradition, but a means of encouraging precisely the opposite: the fulfilment of individual aspirations within the collective realm. It is worth quoting Galton's ideas in their entirety:

> Eugenics belief extends the function of philanthropy to future generations; it renders its action more pervading than hitherto, by dealing with families and societies in their entirety; and it enforces the importance of the marriage covenant by directing serious attention to the probable quality of the future offspring. It sternly forbids all forms of sentimental charity that are harmful to the race, while it eagerly seeks opportunity for acts of personal kindness as some equivalent to the loss of what it forbids. It brings the tie of kinship into prominence, and strongly encourages love and interest in family and race. In brief, eugenics is a virile creed, full of hopefulness, and appealing to many of the noblest feelings of our nature.[29]

What made Galton's definition of eugenics a characteristic document of programmatic modernism was its blend of the pathos of science with the restraint of religion. If the "nineteenth century was a golden age for incubating new dogmas,"[30] eugenics was certainly one of them, resting on the scientific authority of the natural sciences but aiming to forge a biological theology for the future.

The eugenic ideal was represented simultaneously in relation to science and religion.[31] Some eugenicists thus hoped to combine the emerging eugenic philosophy with other modernist messages of change. In the first volume of *The Eugenics Review*, for instance, Maximilian Mügge unhesitatingly connected Francis Galton to Friedrich Nietzsche: if to the former "belongs the honour of founding the *Science of Eugenics*", to the latter "belongs the honour of founding the Religion of Eugenics."[32] An even more categorical statement came from another British eugenicist, Caleb Saleeby, who unambiguously declared that eugenics is "at once a science and a religion, based upon the laws of life, and recognising in them the foundation of society."[33] As a loyal supporter of Galton, Saleeby was a programmatic modernist also in his conviction that eugenics could empower supporters of ideas of national improvement to determine the relative degrees of reciprocity existing between those scientific terminologies purported by theories of biological perfection and the language of spiritual rejuvenation they utilised. "If the struggle towards individual perfection be religious," Saleeby concluded, "so assuredly, is the struggle, less egotistic indeed, towards racial perfection."[34]

The eugenic vocabulary crystallising from these statements embodied a distinctive utopian vision, one that was not merely biological but also religious and historical. As Edgar Schuster, the first Francis Galton Research Fellow in National Eugenics at the University of London, put it:

> the most important task before the apostle of Eugenics is the dissemination of the Eugenic ideal. By this is meant the co-ordinated group of sentiments, aspirations, and desires, based on a right appreciation of moral and social values, which will lead those in whom they are implanted to enter gladly, but wisely, and with a sense of responsibility, on the duties and privileges of marriage and parenthood.[35]

This was a promising attempt to clarify and systematise, one that eugenicists across Europe and the US were to embrace enthusiastically over subsequent decades.

In many ways, the birth of eugenics expressed Galton's concerns with the evolution of British society. In others, eugenics was a template for something that was much more universal. Across the world, groups of physicians, biologists, sociologists and health reformers helped establish eugenics as the ideology of the future, and groups of activists helped define eugenics not only as a body of ideas but also as a scientific community. The International Society for Race Hygiene – established by Alfred Ploetz in 1907, and which Galton honoured as "vice-president" – actively promoted this belief. The Society wanted to advance "scientific, racial, and social biology, including racial and social hygiene," and applied its precepts, first and foremost, to its members,

> who are willing to regulate their own lives in accordance with the motives of the Society – firstly, by earnest efforts to keep themselves in good condition in body and mind; secondly, by pledging themselves to ascertain before marriage, according to the directions of the Society, whether they are fit for it, and if unfit, either to remain unmarried or to refrain from parenthood; thirdly, by promoting the individual and racial well-being of the rising generation.[36]

The extent to which eugenics was slowly spreading beyond the confines of British academia is illustrated by Galton's Herbert Spencer Lecture entitled "Probability, the Foundation of Eugenics." Delivered at the University of Oxford in 1907, it was immediately translated into Hungarian, thus providing a valuable example of the internationalisation of Galton's theories of eugenics.[37] But this lecture is also important for another reason: Galton was preoccupied with more than just reinforcing his support for biometry, a cluster of statistical methods which Galton's favourite student Karl Pearson applied to the study of biological problems (especially evolution and heredity).

An essential corollary of this endorsement was Galton's strong commitment to the cultivation of individual abilities, talents and faculties; but he was equally adamant about the moral and religious issues related to eugenics. He believed that the time for eugenics had not yet come, and that advancing the public opinion's favourable reception thereof was a matter of targeting society at large instead of individuals. In a visionary way, Galton concluded:

> Considering that public opinion is guided by the sense of what best serves the interests of society as a whole, it is reasonable to expect that it will be strongly exerted in favour of Eugenics when a sufficiency of evidence shall have been collected to make the truths on which it rests plain to all. That moment has not yet arrived. Enough is already known to those who have studied the question to leave no doubt in their minds about the general results, but not enough is quantitatively known to justify legislation or other action except in extreme cases. Continued studies will be required for some time to come, and the pace must not be hurried. When the desired fullness of information shall have been acquired then, and not till then, will be the fit moment to proclaim "Jehad," or Holy War against customs and prejudices that impair the physical and moral qualities of our race.[38]

This was the kernel of Galton's eugenic philosophy because public awareness and scientific investigation were not only closely related, but in many ways the entire edifice of practical eugenics rested on as wide a dissemination of knowledge about racial improvement as possible. However, and as so many eugenicists in Europe recognised, there was no simple transition from seeing eugenics as an "academic question" to implementing the "Holy War" of practical eugenics.

The Menace of Degeneration

Dedicating his 1892 modernist indictment of degeneration to the Italian criminologist Cesare Lombroso, Max Nordau reminded him that

> [d]egenerates are not always criminals, prostitutes, anarchists and pronounced lunatics; they are often authors and artists. These, however,

manifest the same mental characteristics, and for the most part the same somatic features, as the members of the above-mentioned anthropological family, who satisfy their unhealthy impulses with the knife of the assassin or the bomb of the dynamiter, instead of with pen and pencil.[39]

Characterising the European fin-de-siècle, and France in particular, in terms of generalised decadence infused modernist epistemologies of displacement and fragmentation with a new sense of direction. Nordau recognised this situation perfectly when he depicted Western civilisation as condemned to extinction. "In our days," he wrote, "there have arisen in more highly-developed minds vague qualms of a Dusk of the Nations, in which all suns and all stars are gradually waning, and mankind with all its institutions and creations is perishing in the midst of a dying world."[40] But, from the vestiges of the old world a new one was to emerge because, as Roger Griffin noted, "an organic process by which degeneration is the prelude to a regeneration in which the old is subsumed within a new form" is intrinsic to the modernist palingenesis.[41]

Nordau's apocalyptic vision echoed the concerns of many eugenicists disturbed by the alleged racial deterioration of the European nations. Just a couple of years earlier, in his acclaimed *Degeneration: A Chapter in Darwinism*, Edwin Ray Lankester presented his evolutionary theory of degeneration, insisting on the parasitic nature of his contemporary society, which he believed – in Darwinist fashion – "may be defined as a gradual change of the structure in which the organism becomes adapted to *less* varied and *less* complex conditions of life."[42] The exaltation of the genetic heritage and the hereditary patrimony became a favourite topic for eugenic propaganda. Some authors outlined degeneration in a variety of historical contexts, universalising it. The Italian anthropologist and eugenicist Giuseppe Sergi, for instance, visualised degeneration as the prevalent human condition, as increasingly more individuals "survived in the struggle for existence," living in "inferior conditions" and parading the physical marks of their "inferiority."[43] Sergi's emphasis on the specificity of degeneration as a signifying modern condition was echoed by eugenicists across Europe.

In 1891, the German physician Wilhelm Schallmayer, one of the founders of German racial hygiene movement, published *Concerning the Imminent Physical Degeneration of Civilised Humanity and the Nationalisation of the Medical Profession*, which Sheila Faith Weiss identified as "the first eugenics tract published in Germany."[44] Schallmayer spelled out the negative effects that modern civilisation had on the organic nature of humanity.[45] Modern science, through medical expertise, scientific knowledge, penal and corrective institutions – in other words, the normative rationality of the state – only contributed to the pathologisation of society. By conceiving degeneration in medical and epistemic terms, and situated within the interstices of power and knowledge, Schallmayer found its modern articulation in the discourse and practice of heredity and eugenics. What this scientific treatment of degeneration marginalised and penalised was the "unfit" and racially "unworthy" individuals seen to be impeding the race's improvement.

Alcoholism is one seminal example of this line of eugenic reasoning. Health and social reformers were divided between those who explained degeneration in terms of inappropriate social environment and those who saw it as caused by hereditary disposition. Eugenicists largely adopted the latter view, specifically encouraging abstinence, and not just for the sake of the individual's health but equally to ensure the racial survival of future generations. At the IXth International Congress against Alcoholism held in Bremen in 1903 both Alfred Ploetz and Ernst Rüdin vehemently condemned the consumption of alcohol, challenging other environmentalist views on the topic and rationalising racial hygiene as the scientific technology most appropriate to protect the individual and the community from the perils of degeneration.[46]

Dwelling on the same uneasy relationship between environment and heredity, Karl Pearson likewise argued in his *The Problems of Practical Eugenics* that "the attempt to improve the racial fitness of the nation by purely *environmental* reforms, the removal of child and mother from unhealthy surroundings and the provision for the weak and the suffering" has failed "in promoting racial efficiency."[47] His diagnosis was indeed pessimistic: "we find ourselves as a race confronted with race suicide; we watch with concern the loss of our former racial stability and national stamina."[48]

But eugenics could provide the antidote to degeneration. "Artificial selection," Sergi noted, "is regeneration."[49] The eugenic tropes of rejuvenation and improvement were used as a counterweight to racial and social degeneration, indicating the overlapping between the protection of the individual body and that of the national community. The Spanish eugenicist Enrique Madrazo had proposed the creation of a Centre for the Promotion of the Race already in 1904, expounding his conviction that "it is in our scope to be able to purify the body and the soul of the human species, so that the black marks which today sadden its existence are removed." Madrazo also underscored the secular scientistic dimension of his eugenic programme. "The religious ideal," he instructed, "must be substituted by that of science; in other words, the sentimental ideal must be substituted by that of reason, which has already replaced the former in those societies which are in the forefront of culture."[50] This intermingling of science and religion encapsulated for Madrazo the new eugenic ideal that would ultimately transform and modernise Spain.

As a cluster of social, biological and cultural ideas, eugenics centred on redefining the individual and its national community according to the scientific laws of natural selection and heredity. In the majority of cases, eugenics relied on the state's intervention to assure the success of its programme of biological rejuvenation. Accordingly, Galton, Ploetz, Madrazo and other contemporary eugenicists strongly believed that the regulatory functions of the modern state ought to ensure not only social control and discipline, but to also provide mechanisms enabling racially "valuable" individuals to flourish and reproduce in greater numbers. Confronted with such difficult circumstances, Pearson, for instance, believed the eugenicists' responsibility to be twofold: they could either "follow the easy course of appeal to popular feeling and untutored human emotion, in which case they will create, like philanthropic effort, immediate interest, have their day and their fashion, and leave no progressive impress on racial evolution." Conversely, they could "take the harder road of first ascertaining the laws which regulate the human herd, of creating a science which shall dictate an ultimate eugenic art."[51]

These anxieties about racial prowess were compounded by the spectre of physical degeneration. In Britain, the Boer War of 1899–1902 brought these fears to the general public's consciousness, adding to a general impression that the British population had began to decline numerically. In fact, social commentators were not far from the truth as, for instance, the birthrate more than halved between 1871 and 1914.[52] Authors like the Fabian socialist Sidney Webb, in his 1907 *The Decline of the Birth-Rate*, and the eugenicist Ethel Elderton, in her 1914 *Report on the English Birth Rate*, confirmed these sombre predictions of social decline.[53] No surprise, then, that an Inter-Departmental Committee on Physical Deterioration was established in 1903, reflecting growing public and official concerns with national degeneration.[54] Such biological predicaments were, however, not reserved for Britain alone.

New discourses on degeneration had emerged in Europe in the wake of the publication of such classics as the 1857 *Treatise on the Moral, Intellectual and Physical Degeneration of the Human Species* by the French psychiatrist Bénédict August Morel, and the 1875 *Heredity: A Psychological Study of Its Phenomena, Laws, Causes, and Consequences* by the French psychologist Théodule Ribot. These discourses, as Daniel Pick has persuasively demonstrated, were built on "various conceptions of atavism, regression, relapse, transgression and decline within a European context so often identified as the quintessential age of evolution, progress, optimism, reform or improvement."[55] If French psychiatry provided one source for this anxiety about degeneration, Italian criminal anthropology provided another. In a number of books including *The Criminal Man* (1875) and *The Delinquent Man* (1876), the criminologist Cesare Lombroso transformed the criminal body into an example of deviancy and degeneration.[56] At the confluence of these intellectual endeavours was the need to identify the individual body as an appropriate object of scientific knowledge and regulation. Eugenicists adopted many of these arguments in their crusade for racial improvement. Increasingly, the race's alleged degeneration became central to visions of national survival amidst mounting international crises.[57]

Some authors such as, for example, the British physician Robert Rentoul insisted in his 1903 *Proposed Sterilisation of Certain Mental and Physical Degenerates,* and again in his 1906 *Race Culture; or, Race Suicide?,* that the "white race" was in peril due to the low birthrate exhibited by the better classes (as opposed to the high birthrate of the lower classes), coupled with racial intermarriage and the increasing number of feeble-minded individuals and asocial elements. "I have avoided entering upon the question of environment," Retoul emphasised, "as a cause for degeneration. Heredity is the great cause."[58] Of the social vocabularies available to eugenicists to articulate their concern with the increased number of "unfit" elements within society, one consistently used was the questioning of government welfare programmes or protective legislation on the grounds that such social reforms enabled the hereditarily unfit to survive, thereby weakening society as a whole. Practical eugenics was hence considered to be the best alternative to social and racial degeneracy. Havelock Ellis shared these views when, commenting on feeblemindedness, he stated that "by specially training the feeble-minded, by confining them in suitable institutions and colonies, and by voluntary sacrifice of procreative power on the part of those who are able to work in the world – we shall be able, even in a single generation, largely to remove one of the most serious and burdensome taints in our civilization, and so mightily work for the regeneration of the race."[59] In Germany, Alfred Ploetz, together with dramatist Gerhard Hauptmann, established a League to Reinvigorate the Race as early as 1879, with the aim of restoring the German nation's racial vigour.[60] A few years before Ploetz published his treatise on racial hygiene Hauptmann described the convolutions of scientism and the new biological morality emerging in Germany at the end of the nineteenth century in his 1889 drama *Before Dawn*, possibly "the first major public statement in Germany on racial biology."[61]

Whether exposed through literature or science, there was, around 1900, an increased anxiety about the nation's physical degeneration. Some authors, like Giuseppe Sergi, transferred the locus of degeneration to a cluster of idiosyncratic historical and cultural entities which he grouped into a new racial body, the "Latin nation."[62] To others, the personification of degeneration was accompanied by the formulation of rhetorical, institutional and disciplinary strategies of social and racial protectionism. To this end, racial decline was consciously politicised. If in Germany debates on degeneration augmented a revival of conservative and right-wing ideologies, in France, some of the most vocal defenders of the race came from the revolutionary Left. The birth control campaigner and anarchist Paul Robin – to name the most conspicuous example – placed the issue of fertility among the lower classes at the centre of his neo-Malthusian programme of racial rejuvenation. If authors like Georges Sorel and Charles Péguy predicated the regeneration of France on patriotism and the Christian tradition,[63] Robin preferred a rational planning approach to motherhood. In 1896 Robin established the League for Human Regeneration together with the journal *Regeneration*, based on a programme of racial improvement he had already outlined in his 1878 *The Sexual Question*, arguably one of the most radical manifestos of anarchist eugenics. Here is a characteristic sample of Robin's scientific management of the human reproduction and its social implications:

> Oh, you who are called proletarians (that is to say, makers of children), you who are crushed by an excess of labour, you who are poorly housed, badly dressed, poorly fed, if you sense your pains, if you aspire to things the possession of which would permit you to struggle against the tyranny so well organised by your oppressors, do not burden yourself with a great number of beings more feeble, more powerful than you! Do not encumber yourself with children! [...] Such prudence is as desirable in the daily industrial battle as it is in the violent struggle of the day, very close at hand I hope, of the social revolution.[64]

Opposing other eugenicists who sought to encourage the poor to limit their reproduction in order to counteract the negative effects of modernity, Robin argued for birth control as a technique through which the poor could, in fact, control the internal dynamic of change in society. As his Spanish disciples, Luis Bulffi, pointed out whilst launching the journal *Health and Strength* in 1904, the purpose of eugenics was

> to make available biological and social scientific data so that future generations are not like our own and so that the ones about to be born are not the result of hurried passions, of a chance sexual encounter. Instead, they should be the result of the conscious decision of healthy parents, who have vigorous bodies and minds, and who are perfectly aware of the task they are undertaking.[65]

The control of reproduction held the key to social and biological regeneration.

That this was clearly the case can be seen in the developments of discourses on degeneracy in Eastern Europe. In 1874, the founder of Romanian psychiatry Alexandru Sutzu established a direct relationship between heredity and the degeneration of nations,[66] whilst in his 1876 *Medical Philosophy: On the Improvement of Human Race* the dermatologist Mihail Petrini-Galatzi went even further in advocating that members of communities suffering from hereditary diseases should be discouraged from reproduction.[67] Other intellectuals in Eastern Europe, like Mladen Jojkić, offered a historical explanation of the current national degeneration, blaming centuries of Ottoman domination for remorselessly encroaching Serbian communities and thwarting their biological development.[68]

The Czech sociologist Břetislav Foustka, on the other hand, integrated the issue of national degeneration within the larger ideal of humanity, providing a sense of a biologically guaranteed physical order and moral purpose.[69] In his 1907 *Abstinence as a Cultural Problem with Particular Consideration to Austrian Nationalities*, Foustka provided a further reading of the idea of resisting decay and degeneration though an extensive programme of rebirth: "Everywhere one calls for regeneration; and not only political, ethical, and religious but also biological regeneration."[70]

Within the emerging hygienic culture of the twentieth century, the "lower classes" and "inferior individuals" found themselves at once rejected and celebrated, decisively transforming not only representations of societies and nations in terms of their racial quality but also the eugenic discourse itself. As Saleeby warned in his 1911 *The Methods of Race-Regeneration*, "[i]t must be remembered that we shall not raise or regenerate the race merely by purging it of diseased elements, however necessary and desirable that process may be."[71] But with few notable exceptions, eugenicists were not interested in redeeming those deemed "unfit," preferring to idealise the healthy body of the nation and the race instead. The internal trajectory eugenics followed might have differed between countries like Britain, France, Germany, Romania and Spain, as scholars of eugenics have repeatedly insisted,[72] but the modern technology of the national body proposed by eugenicists was essentially the same across Europe: the desire to control the population's quality by controlling its reproduction.[73] In this manner, negative eugenic practices like segregation and sterilisation, synonymous with complete submission of individual liberties, could be conjoined with the ideal of a healthy national community.

Internationalising Eugenics

The internationalisation of eugenics permitted its rapid absorption into national and regional contexts. No longer an exclusive idiom, eugenics became part of the general European discussion about the racial future of the nation. As early as 1901, in Transylvania, Heinrich Siegmund published the first eugenic treatise in Eastern Europe, enrolling racial hygiene in the service of important Saxon institutions of control, notably the church and the family.[74] In 1905 the first professorial chair in eugenics was inaugurated at University College London, based on Galton's intention "to forward the exact study of what may be called *National Eugenics*." This concept was defined as "the influences that are socially controllable, on which the *status* of the nation depends. These are of two classes: (1) those which affect the race itself and (2) those which affect its health."[75] Eugenics was, of course, open to various interpretations. Thus, and although Galton and his school emphasised the social context, other readings insisted on racial strength (for example German and Scandinavian racial hygiene) and adaptation to a hostile environment (for example French puericulture). All of these readings of eugenics were possible, even within the same modernist narrative of national improvement, and many authors highlighted these difficulties of interpretation.

In the first issue of the *Journal of Racial and Social Biology*, Alfred Ploetz endeavoured to establish racial hygiene as a discipline within its own right,[76] rather than being a mere sub-discipline of social hygiene as the influential social hygienist Alfred Grotjahn had maintained.[77] Racial hygiene, according to Ploetz, Schallmayer and others around them, was exclusively concerned with the hereditary qualities of the population, and its aims were twofold: to increase and further those hereditarily "superior" individuals, and to decrease – if elimination was not possible – those considered racially undesirable. Contrary to social hygiene, which focused on the protection of *existing* hereditary qualities, racial hygiene was *future* oriented as its driving force was towards building a new racial community.

In 1908 another prominent figure of European eugenics, the Norwegian Jon Alfred Mjøen divided racial hygiene into three thematic clusters. The first dealt with negative "measures for diminishing undesirable racial elements," namely "permanent segregation of recidivists in working colonies and the sterilization of the unfit;" the second cluster referred to positive "measures aimed at the increase of valuable racial elements," including "selective internal colonization with schemes for diminishing the movement from country to town;" "introduction of human biology in school and university curricula" and "centrally controlled propaganda in knowledge of the renewal, health and nutrition of the population, with bureaux for giving information on questions of racial hygiene." Finally, the third cluster, which Mjøen termed "prophylactic race hygiene (protection of the unborn child)," consisted of campaigns "against racial poisons, venereal diseases, narcotics, etc.; certificates of health before marriage, including the discouragement of marriage with widely unrelated races; biological assessment of the whole population; immigration control based on biological standards, with powers to prevent admission."[78]

These definitional endeavours emerged in a context of creative institutionalisation and professionalisation of eugenics. Encouraged by the reception of Galton's ideas in Britain, the Eugenics Education Society was established in 1907 in London. Other countries followed this example and in Scandinavia, for instance, a Swedish Society for Race Hygiene was formed in Stockholm in 1909. American eugenicists had established the Station for Experimental Evolution at Cold Spring Harbor as early as 1904, followed by the Eugenics Record Office in 1910. These initial efforts to popularise eugenics in Europe and the US did produce the outcome Galton had anticipated, namely the recognition of eugenics as "a subject whose practical development deserves serious consideration." The French Eugenics Society was established in 1912; in 1913, the Czechs founded "an institution for research in eugenics" and the Viennese Sociological Society established a section on Social Biology and Eugenics,[79] followed shortly thereafter by the Hungarian Eugenics Society in 1914.[80] Also in 1914, an Italian Eugenics Committee convened at the Roman Society of Anthropology.[81] By then, the German Society for Racial Hygiene, established in 1905, was intensely involved in disseminating eugenics to the general public, both at home and abroad, as illustrated by the Hygiene Exhibition organised in Dresden in 1911.[82]

It is on this occasion that Ploetz defined "qualitative" and "quantitative" racial hygiene. In short, "qualitative" racial hygiene focused on: a) the birth rate, both that of the general population and individual mothers; b) the death rate; and c) the birth excess in relation to the racial struggle for existence. "Quantitative" eugenics, on the other hand, was preoccupied with: a) selection, itself divided into 1) non-selective elimination through death or infertility; 2) selective elimination and the elimination of the inefficient from the race; 3) counter-selective elimination through, for example, war; and 4) counter-selective selection such as marriage between blind people). The other components of "quantitative eugenics" included b) reproductive hygiene (Galton's eugenics was considered as belonging to this category); c) individual racial care in relation to reproductive strengths and, finally, d) the racial care of physical and mental abilities.[83]

Expressing the wide geographical diffusion of eugenic ideas was another conference on eugenics organised in 1911 in Budapest by the radical intellectuals grouped in the Sociological Society and publicised in their journal *Twentieth Century*. The conference was as much scientific as it was socio-political, illustrating more than just the ambition of a new generation of Hungarian intellectuals to come to terms with new developments in the social and natural sciences before the First World War. Eugenics was seen as a mechanism capable of decoding social and national predicaments that – according to the organisers – impeded the progress of Hungarian state and society. The debate thus has a double significance: on the one hand, it gave supporters of eugenics in Hungary the necessary opportunity to synthesise their views on heredity and articulate common programmes of hygiene and racial hygiene; on the other, it added a new dimension to general discussions on social and political transformation which characterised the evolution of political thinking in early twentieth-century Hungary.[84]

József Madzsar set the theoretical framework of the conference with his 1910 article "Practical Eugenics."[85] Hereditary factors, Madzsar argued, were paramount to the creation of a healthy individual, and biometry could help explain how certain hereditary traits transmitted from one generation to another and differed in their range of variability. Madzsar was even more critical of approaches to hygienic improvement and social corrections attempted by state and social institutions, and which he saw as working against "the goals of natural selection." Madzsar's eugenic concepts opposed other visions of public assistance and medical reforms based on humanitarian principles. "The present form of social charity," Madzsar claimed, "is even more dangerous because in most cases it obstructs the perishing of elements which are most burdensome and dangerous for society and it encourages their proliferation."[86]

Madzsar's biological radicalism went well beyond simply criticising policies of social charity, especially when suggesting radical policies of negative eugenics including sterilisation. Invoking Plato's argument that the "disabled should be banished from the state," Madzsar maintained the necessity of reverting to this practice in contemporary society. Eugenicists, he continued, should resist their "pseudo-humanism," and "at least pursue the goal of preventing the proliferation of the unfit and promoting the proliferation of the fit." In such a context, the state was invested with biological prerogatives.

Indeed, according to Madzsar, "[i]f the state has the right to deprive citizens of their freedom, of their life even, it undoubtedly has the right to sterilise as well, especially when this can be executed without any other unpleasant consequences for the individual."[87] Ultimately, Madzsar demanded the modern state aim to purify its racial body whilst eugenically controlling the population.

According to another participant, the biologist István Apáthy, degeneration was primarily an innate characteristic rather than the effect of external causes: "The process known as racial degeneration is actually a malady of the species, a malady which can be cured by extremely special methods." Well versed in hereditarian theories, and a supporter of schemes for preventive hygiene, Apáthy adopted both as sources of inspiration for the "special methods" of combating the "malady" of modern degeneration. Thus, he continued, "[p]ublic hygiene is concerned with the improvement of public life conditions and public health; racial hygiene fights against certain maladies which endanger not only the survival of isolated individuals but the survival of the entire species." The science of hygiene, therefore, needed to adapt to the challenges posed by eugenics, especially to the notion that benevolent hygienic schemes resulted in the multiplication of the "unfit." The importance of eugenics for improving hygiene was thus perceived to be substantial. As Apáthy explained:

> The endeavours of these two sciences are in many aspects similar; furthermore, the improvement of public hygiene itself is one of the methods employed by racial hygiene. Yet the latter has also adopted methods which originated in sociology, on the one hand, and in ethics, on the other; and, finally, it adopted an entirely specific biological method as well. This method is the deliberate selection of those elements that protect the race and the impediment of the reproduction of certain individuals who might have a damaging effect on the future generation.[88]

Like Madzsar, Apáthy also realised that the key to realising "eugenic utopia" lay at the confluence between the interests of the individual and the powers invested in the state. The eugenic policy advocated by Apáthy outlined new priorities for social hygiene, and included these within his vision of state welfare. Biological degeneration and images of deteriorating social conditions only strengthened his conviction that the private sphere ought not to be encircled by excessive individualism, but – when the protection of the race was required – be authorised to endorse the monitoring of the biological capital of the nation by the state. "It is a great mistake," Apáthy noted, "to believe that the interest of the individual is in conflict with state interest, or that private interest fights public interest in the problems of eugenics."[89] He was not the only eugenicist to think in these terms.

In a lecture prepared for the 1912 International Congress of Eugenics in London, the Czech eugenicist Ladislav Haškovec described the "modern eugenic movement" as an endeavour to apply to society "the new biological discoveries of human heredity," arguing equally that new systems of public health and public hygiene were needed to accompany modern sanitary reforms. It was important, Haškovec emphasised, that eugenics focused not only on the health of the individual but considered health and medicine as factors in shaping a higher social condition. To illustrate his arguments, Haškovec pointedly discussed the social consequences of the laws of hereditary pathology, in the cases of tuberculosis and syphilis. Defending the need for a eugenic pedagogy, Haškovec ultimately wanted to promote the introduction of new hygienic and medical principles as the accepted basis of the modern approach to marriage and family. Inextricably tied to the health of "future generations, the nation and the state," eugenic marriage certificates certified both the struggle against social degeneration and social deviance, as well as the fusion of science and national progressivism.[90]

Plans for a new social and national order based on scientific knowledge, and in accordance with eugenic principles, were constantly enunciated during these public debates and lectures on racial hygiene and eugenics in Germany and Austria-Hungary. Eugenicists like Ploetz, Madzsar, Apáthy and Haškovec took a keen interest in contemporary social problems, and their various perceptions of eugenics emphasised the biological rejuvenation of society that was inherent in philosophies of social progress. Some of the assumptions formulated during these conferences were based on local, German, Hungarian and Czech experiences; others were drawn from similar debates in other European countries, particularly in Britain. Karl Pearson's synthesised this programme of "national eugenics" thus:

> Every nation has in certain sense its own study of eugenics, and what is true of one nation is not necessarily true of the second. The ranges of thought and of habit are so diverse among nations that what might be at

once or in a short time under the social control of one nation, would be practically impossible to control in a second. Eugenics must from this aspect be essentially national, and eugenics as a practical policy will vary widely according as you deal with Frenchmen or Japanese, with Englishmen or Jews.[91]

This interpretation of "national eugenics" encapsulates the objectives set by the organisers of the 1911 Dresden Exhibition and the conference on eugenics in Hungary. It was in this context that participants found the necessary concepts to identity the social and medical problems afflicting their societies, and which they hoped to solve through the development of a specifically German and Hungarian form of eugenics. Again, Pearson must be invoked to provide the essential ingredient necessary for these visions of national eugenics to emerge as

> within ten to fifteen years national eugenics will be everywhere a branch of academic training, and that in less than twenty years legislators will accept the fundamental results of the science of eugenics as indisputable facts. What is more, the nation that favours these studies most heartily and most readily accepts the knowledge gained as a guide to practical conduct is destined to be the predominant state of the future.[92]

Yet these attempts at formulating a national eugenics tailored to reflect the local realities of each nation did not preclude the ambition to formulate an international movement of ideas. By the time the First International Eugenics Congress convened in London in 1912, Galton's first commandment – the popularization of eugenics "as an academic question" – had been embraced by more than 400 participants from most European countries and the US. In his presidential address Leonard Darwin, the Chairman of the Eugenics Education Society, phrased the hopes of the eugenicists along the following lines:

> The struggle may be long and the disappointments may be many. But we have seen how the long fight against ignorance ended with the triumphant acceptance of the principle of evolution in the nineteenth century. Eugenics is but the practical application of that principle, and may we not hope that the twentieth century will, in like manner, be known in future as the century when the Eugenic ideal was accepted as part of the creed of civilisation? It is with the object of ensuring the realisation of this hope that this Congress is assembled here today.[93]

The concert of eugenic voices in attendance was far from homogeneous. Italian eugenicists like Corrado Gini and Giuseppe Sergi insisted on coupling eugenics with demography and natality; French eugenicists like Frederic Houssay and Adolphe Pinard criticised the inclination towards negative eugenic practices like sterilisation showed by some participants, especially from the US, and advanced their version of puericulture instead,[94] whilst the German eugenicists like Agnes Blum and Alfred Ploetz popularised their concept of race hygiene.[95] Rather than being satisfied, Ploetz looked to the future, a time when a new hygienic and racial order would be instituted. By thinking in terms of the collective racial community rather than the individual Ploetz aimed to create a eugenic programme ideally suited to a culturally and biologically rejuvenated society:

> We must instil in our children greater courage to undertake the responsibility of life, a higher patriotism, a sense of devotion to our race which must face the great combat of future, so that they gladly prepare for an expenditure of energy beyond their own immediate and personal interest. Only then, in the decision between self-centred individualism and service toward a new generation and new forces for their race, will they decide in favour of life.[96]

American eugenicists such as Bleeker van Wagenen, the chairman of the Committee of the Eugenic Section of the American Breeder's Association, discussed the practical application of eugenics. In fact, the first sterilisation law had been introduced by the US state of Indiana as early as 1907, one targeting "undesirable" individuals, especially those deemed to be physically disabled, mentally ill, and criminal.[97] It became clear that eugenicists were not simply conjuring up themes of racial and social decay.[98] Eugenic redemption was gradually explained in terms that overcame the contrast between decadence and modernity to propose a regenerative political biology, one centred on selective breeding and national purification. Thus American eugenics combined programmatic modernism with a certain vitality and rationality, adding as much a spiritual as a collective dimension to national regeneration that other eugenic movements – apart from German racial hygiene – never entirely embraced.

Britain, the US and Germany were instrumental in establishing eugenics as a topic of scientific debate in Europe during the first decade of the twentieth century. Certainly, at the time, supporters and detractors alike praised Galton for his commitment to practical eugenic programmes of social and national rejuvenation derived from theories of evolution and heredity. The emergence of eugenics in other European countries, to be sure, must be connected to specific historical circumstances, including economic growth, class stratification, colonialism and racism; but also, vitally, to the wider acceptance of theories of heredity by the scientific and political community as well as remarkable institutional networking.[99]

As we shall see in the following chapters, with the outbreak of the First World War, and especially during the interwar period, eugenics became part of a nationalist culture, increasingly alienated from the scientific realms within which it had originated. Coupled with the increased biologisation of national belonging eugenics transcended the field of medicine and biology; it gradually operated within a new nationalist register, one unifying the physicality of the nation with its resurrected spirituality. Subscribing to this axiom, eugenics redefined the body politic according to the scientific standards of the age, whereby the nation's physical and spiritual qualities were placed under close inspection by both state agencies and individuals entrusted with the role of protecting them.

2

WAR: THE WORLD'S ONLY HYGIENE, 1914–1918

"The Great War was a eugenics nightmare"[1] considering the astounding casualties. Germany lost nearly two million people, followed by Russia with an almost equal number and by France with approximately a million and a half casualties.[2] A military conflict of such magnitude was certainly beyond the wildest imagination of those who, prior to 1914, glorified war as a means to revolutionise the stale condition of modernity. If the literary and artistic avant-garde depicted war as a therapeutic response to a long process of cultural malaise and degeneration, most politicians believed the time had come for a rearrangement of Europe's political order. All groups, however, deemed the existing European system of power and political alliances as unable to cope with the new forms of nationalism, imperialism and political anarchism that arose in the early twentieth century. Take, for instance, the Italian poet and founder of the Futurist avant-garde, Filippo Marinetti who famously portrayed war as "the world's only hygiene" in 1909.[3] Modernists believed that when all cultural values had proven themselves inadequate, the rejuvenation of the national community could only result from the existing order's total transformation. Their longing for a spiritual renewal grounded in the hope that the nation would ultimately be saved by the violent transgression of all existing boundaries, as expressed in Marinetti's address to the Italian students: "[o]ur ultra-violet, anti-clerical, and anti-traditionalist nationalism is based on the inexhaustible vitality of Italian blood and the struggle against the ancestor-cult, which far from cementing the race, makes it anaemic and putrid."[4]

War was seen as the "Gründerzeit," the founding historical moment when empires disappeared and new nations were forged from the vortex of universal barbarism. According to the Spanish eugenicist Antonio Vallejo-Nágera, during the war, "[r]aces which have rediscovered themselves, which have contemplated their history, the peoples which have struggled to recuperate their spiritual values and revive ancient traditions, [were] those, which, like a phoenix, have been reborn from the ashes and have been able to stand up the whole world in order to maintain their racial personality."[5] In this rejuvenating discourse, war symbolised a new beginning; it transformed the race as much as it changed the life of ordinary individuals. It was a process of "self-transcendence through war,"[6] one which Emilio Gentile described as metanoia, the transformation of "the old man into the fighter or the new man."[7]

But war provided more than just a mythology of violence and a matrix for the sacralisation of the nation; many authors around the turn of the twentieth century saw it in terms of a redemptive return to a biologically superior condition, akin to the project of human improvement prophesised by the eugenicists. It was, in fact, during the war that eugenics first extended its reach beyond scholarly debates and infiltrated public and political discourses, confirming Kevin Repp's comment that "[s]cience seemed poised to come to the rescue of culture in the summer of 1914."[8] Some eugenicists deemed war to be the pinnacle of human evolution, a mechanism through which to regulate over-population and the demand for economic resources. As the rhetoric of biological pessimism grew throughout Europe and the US around 1900, these authors asserted that war was the ultimate illustration of national potency; war was effective in mobilising the racial abilities of the nation whilst simultaneously counteracting physical degeneration and racial miscegenation; war, ultimately, proved the superiority of one race over another in the perpetual struggle for survival. Following the distressing impact the Boer War in South Africa had on the British perception of national health, Karl Pearson, for example, notably commended the positive functions of war and its redeeming features of sacrifice and racial purification.[9]

Moreover, war encouraged many eugenicists to attempt the necessary codification of eugenics as they vociferously argued that the state should control biological reproduction in the interest of national efficiency and to counteract the effects of war. Theoretic and practical eugenic themes became increasingly intertwined as control of the nation's body politic brought the eugenicists in direct contact with the political sphere. But eugenicists were divided over the issue of whether war itself was beneficial or detrimental to social and national progress. Some argued that war perfectly illustrated processes of natural selection in social and national spheres, whilst others argued that war led to the elimination of the physically and psychologically healthy, thus allowing the "weak" to increase in number and subsequently determine the biological destiny of future generations. The British eugenicist J. A. Lindsay logically opined that "[t]wo theories are possible regarding the eugenic and social influence of war in general," where in the first theory "war is, in the main, profoundly dysgenic and anti-social, wasteful of the best life of nations, destructive of capital and of the fruits of industry, a propagator of disease, hurtful to the stock, a well-spring of international hatred and alienation." Alternatively, a second line of reasoning argued that "war is a tonic, though admittedly a severe tonic, to the nations; that it promotes the virile virtues – courage, endurance, self-sacrifice; that it imposes a wholesome discipline; that it is a great school of patriotism, efficiency, and national solidarity; that prolonged peace leads to softness of manners and racial decadence."[10] In this chapter, we will take a closer look at some of these eugenic theories to illuminate how eugenics hoped to combat the alleged racial erosion of the national community caused by the war.

War as an Educator

The outbreak of the First World War in 1914 undoubtedly marked a crucial shift in eugenic rhetoric and practice. Eugenicists everywhere invoked themes of racial loss and disrupted birthrates, but, more significantly, visualised these in terms of national virility and youthful sacrifice. "Eugenics and war – the clash between ideals," the Scottish naturalist J. Arthur Thompson remarked sombrely. The eugenic rhetoric of national survival was invoked to describe the negative impact war had on the race. From this standpoint, Thomson validated a biological interpretation of war. "Let us not seek to conceal the fact," he wrote, "that war, *biologically regarded*, means wastage and a reversal of eugenic or rational selection, since it prunes off a disproportionately large number of those whom the race can least afford to lose."[11]

When analysed culturally and spiritually, however, Thompson was hopeful that war would contribute to "a heightening of the standard of all-round fitness," as "forms of national weakness" will be resisted and "the tare seeds in our inheritance" prevented from "germinating." Moreover, war was "a time of vivid national self-consciousness and of freshened idealism," which Thomson placed at the centre of his vision of a regenerated future British Empire. The transformation of each individual within the body politic of the nation was one immediate consequence of the war: "We are going to know and to like one another better, having fought together, rejoiced and sorrowed together; we are going to see more of one another as distance-annihilating devices increase and cheapen."[12]

Thompson used the unique conditions created by war as the locus for the much needed eugenic palingenesis of society. "As eugenists," he remarked, "we are concerned with the natural inheritance and its nature, which is fundamental, as men we are also concerned with our social heritage, which is supreme." The decidedly modern function of eugenics as a source of biological renewal and sustenance meant that war – its devastating effects notwithstanding – would create the favourable conditions for new national values to emerge. "We cannot end without expressing the hope," Thompson concluded, "that even if the natural inheritance in our race must suffer impoverishment through the tragic sifting of this terrible war, we shall win through in the end with our social heritage enriched."[13]

Thomson was not the only one to mythologise war as a metaphor for national regeneration. "War is a great educator," concurred J. A. Lindsay. The anomie of the modern age could be overcome, he believed, as war "tends to enlarge our vision, to soften our prejudices, to mitigate our self-sufficiency, to moderate that insularity of mind which has so often, and not quite causelessly, been charged against us as one of the foremost of our national failings."[14] The perception of war as the educator of the national spirit was shared by eugenicists in other European countries as well. The Hungarian biologist Lajos Méhely invoked war to foster nationalism, and praised it for the biological rejuvenation it entailed. Patriotism, Méhely also believed, was retreating in a modern society overwhelmed by decadence and contemptuous of discipline and self-sacrifice. A new generation of nationalists was being formed in the trenches, he believed, one that was racially healthy and willing to sacrifice itself in the service of a greater cause. Emilio Gentile has offered a potent interpretation of this view, arguing that, in the case of Italy, the "life of the trenches, the sense of comradery, the soldier's reciprocal dependence and loyalty in battle became the basis of a new sentiment of national communion, imbued with lay religiosity."[15] War, in other words, was an opportunity for both spiritual and physical renewal, an antidote to degeneration and the chance to enhance the nation's biological qualities. As Méhely insisted, natural selection prompted by war would eliminate "the pale, the weak, the nervous and, from a military point of view, a simply valueless generation."[16] The result will, according to Méhely, be "the breeding of patriotic, strong-willed, disciplined and bodily strong generations of citizens."[17] An essential corollary of this strong commitment to the cultivation of individual abilities and faculties was the process of selective breeding numerous eugenicists advocated.

A "constructive eugenic policy in time of war," according to the British eugenicist Theodore Chambers, must concentrate on both those actively engaged in warfare, such as the soldiers, and those left behind, at home, namely their wives and families. By substituting war with eugenics in Galton's definition of the term – "War is shown to be an agency, under social control, which tends to impair the racial qualities of future generations physically and mentally"[18] – Chambers separated war from its gruesome context, and proclaimed it to be a metaphysical force for national regeneration: "War may be glorious in prospect. It may be inevitable. It may be justifiable. While it lasts we may suffer, but the excitement is intense. Every nerve is strained to bring to it but one conclusion: victory."[19]

For these authors war was the instrument of collective social catharsis, the greatest of tests of the nation's racial fitness and morality, and as such could be rationalised as the educator of the national spirit. The centrality of national rebirth ensured that eugenic visions of racial improvement operated across cultural, ideological and ethnic divisions. Commenting on the Jews fighting for the German army, Todd Presner remarked that "[e]mboldened by their heroic fighting tradition and physically regenerated in the gymnastic-halls of modern-day Europe, Jews would bravely serve the German fatherland and prove, once and for all, that they were a muscular, military people."[20] The German Jewish journalist Binjamin Segel exemplified this attitude in his 1914 article "The War as Master Teacher," claiming that "[u]nlike any other historical event war answers the question of how much bravery, contempt for death, discipline, organizational capacity, sacrificial courage, and physical strengths lies within a people."[21] By participating in war, Segel believed, qualitative ethnic differences between Jews and Gentiles have at last become de-essentialised. Patriotism transgressed the racial and corporeal essence between German nationalism and Zionism, as it was allegiance to the country that ultimately mattered. The war thus occasioned Jewish leaders to equate national consciousness with political assimilation, accentuating the need for collective palingenesis. It was, to quote the British suffragette May Sinclair, the "Great War of Redemption."[22] When referring to the neighbouring Austro-Hungarian Monarchy, István Deák similarly argued that "World War I marked the apogee of Jewish participation in the life of Central Europeans. In the delirious enthusiasm of August 1914, Jews were amongst the greatest enthusiasts. They endorsed the war, in part because the enemy was the anti-Semitic Russian Empire, in part because the conflict's outcome promised to bring their final and complete acceptance."[23] War, ultimately, exemplified how a degenerative present could be overcome through physical hardship and military combat, whilst offering new models of social and political renewal that would allow the nation as a whole to fulfil the task of building new foundations for the future.

However, this is just one aspect pertaining to the relationship between eugenic theories of human improvement and war. Concomitantly, questions about the national body's deteriorating health at the hands of war became more articulate. As noted in the previous chapter, eugenics operated within a conceptual framework that allowed for the possibility of systematically classifying the nation's social composition according to racial strengths and their cumulative impact on future generations. Due to this theoretical malleability, there was a concurrent emphasis on social, biological, and cultural categories upon which the health of the nation could be modelled. Thus, there were many eugenic voices arguing against views depicting war as the perfect gardener pruning the world through natural selection and the survival of the fittest. For this category of eugenicists, war was the symbol of racial destruction and the perfect mechanism facilitating the weakening of the social fabric of society.

To be sure, the healthy and strong race so often invoked by eugenicists was an ideal to which many of those supporting war adhered. But the task of realising this ideal was managed differently in European countries and cultures. War was complexly mediated by the local variables shaping each country involved in the conflict; but it was also eugenically conditioned by a generalised political biology, one defined by the traumas, displacements and tragic realities of military conflict. Universally, eugenicists berated the state for failing to recognise its biological duty to assist those sections of the nation it deemed a source of racial wealth, namely to encourage them to reproduce through social assistance and charitable work. The Italian anthropologist Giuseppe Sergi openly called upon the state "not only to maintain the high spirits of the nation and the power of resistance to the harsh conditions of the war, but also to maintain healthy and vigorous bodies for the present and the future."[24] Thus, war offered the state the possibility to control the nation's health and identity. But with the number of military casualties rapidly increasing, the role of the state – so the eugenicists maintained – should not only be one of patrolling the racial borders of the nation but also one of protecting the internal racial disintegration caused by war.

Nations Racially Damaged

Within this context, eugenic reasoning sought to address the impact of military and civilian causalities in relation to hereditarian laws, and more broadly, in the terms of Darwinist doctrine of evolution. Discussing how natural selection affected the life of nations, Charles Darwin lamented the counter-selective consequences of war already in his 1871 *The Descent of Man*:

> In every country in which a large standing army is kept up, the finest young men are taken by the conscription or are enlisted. They are thus exposed to early death during the war, are often tempted into vice, and are prevented from marrying during the prime of their life. On the other hand, the shorter and feebler men, with poorer constitutions, are left at home, and consequently have a much better change of marrying and propagating their kind.[25]

Alfred Ploetz had similarly pointed to the dysgenic effects of war in his 1895 treatise on racial hygiene, and subsequently proposed that the "worst individuals" be drafted into military service in order for "healthy individuals to be saved."[26] A characteristic proposition of this view that war was ultimately detrimental to the biological improvement of human societies is offered by an anonymous contribution to *The Lancet* in 1916, in which the author bluntly declared:

> War is antagonistic to eugenics. It is not like the storm that uproots parasitic plants, the hurricane which purifies the atmosphere, but is the whirlwind that shatters the forest trees, beats down the corn and devastates the fields where nettles will afterwards flourish. War is not a reviving blood-letting, but an exhausting haemorrhage, which blanches the life-producing organs and prepares the soil for the development of pathogenic germs. War scatters the seeds of disease, sorrow and hate, and death, all of which cause the deterioration of the race.[27]

War was not just the epitome of national resurrection, but also a mechanism of eugenic action, demonstrating a clear affinity between eugenicists and other groups concerned with social and population questions, including neo-Malthusians, feminists, social hygienists and health reformers. Although their prescriptions were often different, especially with respect to the issues of reproduction and fertility, they all appeared to share a number of similar concepts and approaches, notably the belief that the health of the community was both a biological and moral issue.

With the intensification of the military conflicts in 1915 and 1916, following the Dardanelles campaign and the battles on the Eastern front and Verdun, it became obvious that war inflicted severe social, demographic, and eugenic damage upon all nations involved. This explosion of militarism was often accompanied by racial prejudice and stereotyping.[28] Concerned eugenicists were often enlisted, or volunteered, to demystify such practices. Yet, these efforts were powerless. As the Swiss eugenicist Max von Gruber remarked; "[w]hat we [the Germans] believe to be the illogical predilection to pathetic idle talk, the lack of the love for truth, and hysterical excitability is perceived by [the French] as the beautiful power of the inspired souls that overcomes all the earthly burdens."[29]

French authors, on the other hand, consistently negated German claims of racial and cultural superiority, insisting – like the anthropologist George Poisson – that miscegenation and a succession of historical migrations had irremediably dissolved the "pure" Aryan and Nordic nature of German racial type.[30] Louis Capitan, a professor of pathology at the Collège de France, went even further and connected a eugenic definition of inferiority to pathology and morbidity. German and Austrian soldiers were, according to Capitan, in vast proportions affected by criminal degeneracy and should consequentially be regarded as morally irresponsible. In fact, Capitan believed that the Austrian and German armies were, in fact, trying to overcome their sense of degeneracy and inferiority by prolonging the war, thus providing official propaganda with eugenic idioms.[31]

These allegations of racial deficiency illustrated not only a radicalisation of war, but also the emergence of a new eugenic language to accommodate it. Eugenicists had many ideologically persuasive sources with which to outline their new programmes of racial and national survival. These sources consisted not only of detailed and striking clinical descriptions of the enemy's pathological body like Capitan's, but also of aggressively promoted demographic and medical statistics concerning what was perceived to be a decrease in the racial quality of the population. According to one commentator, "whatever argument may be adduced in favour of the preparations for war as of eugenic value in our times, it must be generally conceded that war itself, under modern conditions of mechanics and mobility, is almost entirely dysgenic."[32]

There is an important corollary of this view. The assignment of eugenic significance to war implied that the realities of military conflict, and its ensuing consequences, invariably also confirmed the presuppositions advanced by eugenicists. Most Italian eugenicists, for instance, recommended a negative hermeneutics of war. In his 1917 *War and Population* the demographer Franco Savorgnan offered a convincing description of the qualitative and quantitative decline of "the racial type" of Italian men, and their reproductive capacities, due to their participation in the war. "The great majority will be," Savorgnan believed, "undermined by privations, venereal diseases and tuberculosis, or, in the best hypothesis, will have brought home from the war a nervous system strongly prejudiced by the ceaseless fire of the artillery."[33] It was thus realised that eugenics ought to have an impact upon society at the national level, and that eugenicists were not merely messengers of science, but guardians of the nation's biological treasure. That they – and, by extension, the medical sciences – could treat and heal the wounds of the nation.

But eugenic readings of war changed according to author, and the racial meanings attached to the nation were not fixed. The process of interpreting the dysgenic impact on the population was consequently not a process of assigning fixed meanings to war but, rather, that of reading the patters of structured modification of the racial quality of the population. The Italian statistician Corrado Gini was, illustratively, less inclined to wholeheartedly accept pessimist visions of racial destruction purported by war. Arguing that "chances of destruction were better equalized amongst individuals of relatively greater or less value," Gini argued that one should look at the positive eugenic consequences of war. "In practical eugenics," he contended, "the important thing is not to have a fixed eugenic ideal, which can seldom be realized, but to have adequate criteria for discriminating between favourable and unfavourable eugenic factors."[34] If some eugenicists were inclined to see the positive effects of war in its early stages, especially with regards to fostering patriotism and national solidarity, by the end of 1916 a consensus emerged that war was dysgenic as it affected both the combatants and the civilian populations.

Repudiating the restricted usage of biology by politicians and intellectuals, eugenicists now embraced a broad racial protectionism. Leonard Darwin synthesised the main tenets of the new eugenic politics needed during the war in his 1916 lecture "On the Statistical Enquiries Needed after the War in Connection with Eugenics." Four questions preoccupied Darwin in this text: "(1) At the conclusion of the war to what extent will the nation have been racially damaged? (2) To what extent will this racial damage reappear in subsequent generations? (3) In what ways will this racial damage injuriously affect the nation as a whole? (4) And how can this damage best be remedied?"[35] Darwin viewed the nation as a living organism in which the family constituted a nucleus, and future generations figured as an index of the nation's overall vitality and prowess. Eugenically, Darwin explained that the racial damage "must be measured not only with reference to the qualities of the individual killed, but also with regard to the extent to which those individuals would actually have transmitted these qualities had they lived."[36] In other words, the racial damage depended on whether the soldiers killed were "on the whole above – or below – the mean level of the whole community in what Galton called civic worth."[37] Nonetheless, Darwin acknowledged that acquiring the necessary information for a thorough analysis of these patterns was difficult, if not impossible.

With respect to the second question, Darwin argued that "[i]f the nation will be physically and mentally damaged by the war, these evil effects will certainly affect future generations because of the check thus *directly* produced on intellectual and material progress."[38] Darwin was aware of the declining birthrates and spreading poverty in addition to the economic difficulties caused by food shortages. An important question persisted, nevertheless. If the environment played no significant role in shaping the biological future of the nation, how could one quantify and interpret the casualties war inflicted on the national body? "Our inborn qualities being entirely unalterable," surmised Darwin, "what can eugenic reform, or the attempt to improve these inborn qualities, do [...], it may be asked."[39]

As all concerned soon came to realize after 1914, the war had not only reconfigured national politics and international relations, but afforded the various contemporary theories of heredity a new significance. Where some of his compatriots glorified war's beneficial influence on the race, the Hungarian eugenicist Géza von Hoffmann discussed its detrimental effects. Drawing upon rising fears of racial degeneration and declining popular health, Hoffmann offered provocative evaluations of the consequences resulting from drafting the community's healthy members. Echoing Leonard Darwin, one eugenic trope in particular was used most insistently by Hoffmann in his description of the dysgenic effects of war: the biological capital of future generations. Protecting the racial quality of future generations constituted one of the key forces driving both quantitative and qualitative eugenic policies:

> The aim of eugenics is the greatest possible number and the best possible quality of people. Quantitative eugenics deals with numeric increase whilst qualitative eugenics tries to improve individual value. Having in mind the main laws of heredity, our objective should be to secure many descendants for outstanding individuals and few for individuals below the average. Quantitative and qualitative eugenics are practically impossible to separate.[40]

These arguments were synthesised in Hoffmann's 1916 *War and Racial Hygiene*. His was a pessimist vision in that war, Hoffmann reiterated, amounted to a "total annihilation of peoples."[41] Having described the counter-selective consequences of warfare, he lamented it had also exposed the combatant states to various forms of biological and racial extinction, obstructing population growth and having a dysgenic effect on the hereditary constitution of the European nations.

Using the statutes of the German Society for Racial Hygiene as an example, Hoffmann assessed the "racial burden" and "degeneration" war inflicted on the nation's "genetic stock" towards arguing for the protection of marriage, the introduction of prophylactic measures against venereal diseases and, especially, the discouragement of "inferior people" from reproduction.[42] Hoffmann believed that the "welfare of the race" necessitated an "unrelenting struggle for existence,"[43] as it offered the only way to protect the race from the destructive consequences of war.

This form of race protectionism required eugenics be given the leading role in the process of national recovery, seconded by an equally well articulated population policy. And, as the damages caused by war increased, Hoffmann insisted that broader segments of the political elite and the population should be made aware of these policies' importance. Without a significant improvement of public awareness of these issues, the negative effects of the war would, in turn, result in a profound national crisis.

With Europe facing a difficult present and uncertain future, Hoffmann presented those racial hygiene measures he deemed necessary for "the recovery of the race" after the war. He grouped them into "quantitative and qualitative" categories, with the latter further divided into "positive and negative" forms. Positive racial hygiene promoted the "breeding of those superior," whilst negative eugenics "impeded the reproduction of those inferior."[44] Two countries served as prominent examples, the US and Germany. If the former was the herald of "qualitative" racial hygiene, having already introduced measures to "prevent the reproduction of the inferior," the latter was largely preoccupied with "quantitative" racial hygiene. Only recently, Hoffmann remarked critically, had German eugenicists turned towards "qualitative" racial policies. Such assessments were by no means atypical, but his contemporaries were often confused by the synthetic range of eugenic views on racial preservation and nationalism. Not surprisingly, the German physiologist Georg Friedrich Nicolai attempted to illuminate these contradictory interpretations in his 1917 *The Biology of War*. Not denying the importance of the Darwinian struggle for existence, Nicolai nevertheless perceived war as a destructive plague aggravated by national chauvinism and ideas of racial preservation, especially in Germany.[45]

Within the more specific context of the theme of modernism and war, eugenic ideas of demographic preservation are important because of their inherent eclecticism. The complex relationship between quantitative and qualitative policies conjures up this eclectic eugenic vocabulary, serving to highlight the interface between the public's involvement with eugenics and elitist hereditarian traditions. Especially worthy of note is the distinct prominence given to the crucial process of biological reproduction which broadened the social and political base of eugenics. For those eugenicists, like the Austrian Julius Tandler, anxious to preserve national racial qualities, appeals for a demographic increase were embedded in a plurality of cultural and ideological contexts, and can be adequately understood when set against these discursive backgrounds.[46] Ultimately, one of the important issues raised by eugenicists during this period was about which biological strategy was more appropriate to counter the negative costs of war. In discussing this issue, one needs to reflect on the textual ramifications of eugenics, its embedded proclivity for the practical construction of a new social and national order, one commensurate with the conditions created by war.

The Eugenic Crusade: Quantity or Quality?

The previous section outlined various eugenic interpretations of war, both positive and negative. One must, however, consider the ways in which eugenicists related to, and interpreted, the effectiveness of their theories. As we have already discussed, eugenicists did not have a unified notion of what impact the war would have on the racial qualities of the nation. As a result, two contrasting eugenic conceptions were formulated: namely, the quantity model and the quality model. The notions of quantity and quality were used as heuristic devices by eugenicists, as a means of conceptualising the way in which they perceived their doctrines operating in relation to the racial damage caused by war.

But the acutely alarming descriptions offered by eugenicists did not arise solely from the reality of war, but also from the presumed incongruity of hereditarian theories. Eugenics undoubtedly emerged as one of the most articulate responses to the crises brought about by the war between 1914 and 1918 – but it was an ideology torn between its biological determinism (nature) and social protectionism (nurture). Historians have noted that several prominent eugenicists, among them Galton himself, strongly sympathised with social protectionism and class elitism. As Richard Soloway remarked, the war "brought to a head for the first time serious disagreements about the relative contributions of nature and nurture which were endemic to eugenics and which plagued it throughout its history."[47]

Eugenicists feared both the adverse consequences of declining fertility rates and the equally detrimental effects of the increased number, visibility and fiscal costs of disabled individuals on society. In 1916, Leonard Darwin identified two practices through which "the innate qualities of the nation" were to be preserved and which, if adopted, "would produce racial results of enormous value." The first deemed it "immoral and unpatriotic" for healthy parents to "limit the size of the family"; the second instructed that "parenthood is immoral either when both parents are clearly defective in ways rendering it highly probable that the cause is constitutional, or when one parent is very markedly defective, or when there is not a reasonable prospect of its being possible to maintain the child in mental and physical well being without extraneous help being received by the parents."[48] The mounting human losses and their impact on Britain's demography and social structures were transparent in Darwin's argument, notably in the stipulation that "our nation is now assuredly either advancing or retreating in all the qualities which led to well-being and renown; and to do nothing is not to be stationary, but most probably to drift down the hill of racial decline."[49]

The relentless exposition of these anxieties during the war contributed to the growing prestige eugenics and its solutions to demographic and social crises eventually enjoyed. Nonetheless, these eugenic arguments, relevant as they were for countries afflicted by war, were often couched in a nationalist rhetoric about racial supremacy and survival. In addition to occasioning the introduction of social and medical policies dealing with particular groups, eugenics generated a resurgence of nationalist concerns about the deterioration of the nation's racial qualities. At the inaugural meeting of the Hungarian Eugenics Society in 1914, István Apáthy passionately declared: "We have to start a real eugenic crusade."[50] Like in Britain, Germany, Italy or France, the war constituted the greatest challenge the Hungarian political establishment had faced in its modern history, and eugenics increasingly became part of a new nationalist discourse whose main concern was with the unfavourable effects war would have on the Hungarian nation and state.[51] Other eugenicists in Central Europe shared similar concerns with respect to the impact of war on the health of the population. Describing the aims of the Czech Eugenics Society (established in 1915), the biologist Ladislav Haškovec classified them into:

> (1) the special study of biology; (2) the dissemination of knowledge of the conditions of physical and psychic health amongst all classes of the population, together with the conviction of responsibility toward future generations; (3) the fight against hereditary diseases and those of early infancy; (4) the encouragement of care for women in confinement, of the new-born, and of nursing women; (5) the battle against alcoholism and tuberculosis, against venereal diseases and against all the other factors which destroy the roots of the nation.[52]

Comparably, the French eugenicist Lucien March, considered "alcoholism, tuberculosis, and venereal diseases" to be the "three fatal enemies of eugenics and race hygiene."[53]

There were, however, national particularities, such as different declining fertility rates and depopulation. The German Society for Racial Hygiene expressed its dissatisfaction with the spreading trend towards small or only one-child families, and insisted that a new population policy centred on rural resettlement and opposition to birth-control were urgently needed. "The German Empire," Géza von Hoffmann insisted, "can not in the long run maintain its true nationality and the independence of its development, if it does not begin without delay and with the greatest energy to mould its internal and external politics as well as the whole life of the people in accordance with eugenic principles."[54] Concerns over demography in general, and the quality of the population in particular, were accorded preferential status in the demands made by the German Society for Racial Hygiene. Illustrative examples of these were: "inner colonization (back-to-the-farm movement) with privileges of succession in favour of large families"; "obligatory exchange of certificates of health before marriage"; and "awakening a national mind ready to bring sacrifices, and a sense of duty towards coming generations. Vigorous education of the youth in this sense" was particularly recommended.[55]

These general eugenic prescripts were devised to reflect the German Society for Racial Hygiene's nationalist morality although German eugenicists were far from united in their treatment of war. Two dominant directions crystallised during this period. The first was committed to social and welfare policies. The work on racial hygiene and heredity conducted by the dermatologist Hermann Siemens was particularly important in this respect as it linked population growth to national struggles for competition and resources. Siemens argued that new eugenic policies were needed to strengthen the populations' hereditary protection.[56] Reflecting this trend, the German Society for Population Policy was established in 1915 with the aim of pursuing a quantitative eugenic policy.[57]

The other eugenic direction "saw racial hygiene as a means of national salvation, justifying territorial conquests."[58] For eugenicists like Fritz Lenz and those associated with this current of thought, qualitative measures to increase the worthy elements of the nation were deemed paramount. The League for the Preservation and Increase of German National Strength, also formed in 1915, was an indication of this emerging nationalist eugenics. The gap between quantitative and qualitative eugenic policies also indicated a certain incongruity between the eugenic rhetoric and its translation into everyday practices. This was, in fact, a rupture between various eugenic models.

Radical eugenicists, like Géza von Hoffmann, repeatedly criticised the German eugenics movement for its timidity. He noted, for instance, that German racial hygiene corresponded conceptually to British and American eugenics, although the latter was concerned with the hygiene of the race in a narrow sense. But instead of being preoccupied with how best to protect and improve the race, German racial hygiene focused on social issues like sexually transmitted diseases and alcoholism. Additionally, Hoffmann was concerned that racial hygiene, due to the war's demographic aftermath, was becoming unidirectional and restricted to quantitative campaigns. He strongly opposed this orientation, arguing that qualitative measures to improve worthy and healthy racial elements, and to eliminate the unworthy and the unfit, were essential for a successful eugenic programme.[59]

The German legal expert and founder of the German and Prussian Association for Infant Welfare, Carl von Behr-Pinnow, also contemplated what he thought the most appropriate wartime legislative population policies should be. The intensification of eugenic propaganda efforts in schools and the media was, for example, considered essential towards heightening patriotism and the ambition of reclaiming land from neighbouring countries where disabled soldiers and veterans could be settled after the war.[60] The quantitative model of eugenics pursued by Behr-Pinnow considered the body of the nation to be composed of all individuals irrespective of their alleged hereditary constitution or physical disabilities caused by war. He had thus recognised that even as the quality of the population was being constituted as the main source of social and racial engineering, its hereditarian basis was bound also to suggest a delicate anthropological relativism, one which then led not only to a defence of the quantitative model of eugenics but also to an approving attitude towards those members of the community maimed or disfigured by war. As Seth Koven has suggestively described it, when looking at the "crippled children" and "wounded soldiers" resulting from the Great War in Britain, "the tens of thousands of men who returned home from the battlefronts of World War I permanently disabled, many lacking arms and legs, were dismembered persons in a literal sense but also in a social, economic, political, and sexual sense."[61]

For the eugenicists, the wounded men were, undoubtedly, of racial importance. For this reason, Behr-Pinnow and others argued that no eugenic rejuvenation of the national body could be genuine unless it considered all individuals of the community, whether diagnosed with physical impairments or not. A healthy nation was not merely a matter of building social and charitable institutions from above; it required the cultivation of a sense of racial virtues amongst the population, and this goal could only be achieved through constant moral and biological education. After the war, as we shall see in the next chapter, this idea translated into an array of social and national policies designed to encourage the regeneration of the national body.

Political Biology

Attempts to regulate demographic changes, social predicaments and health patterns during the war were paralleled by institutional activities, especially in the military and health sectors. The popularisation of eugenics benefited greatly from this intensive activity and eugenicists used their official positions in various state departments to implement medical and health reforms. In Germany, according to Sheila Weiss, "[p]rior to the outbreak of hostilities, the government appears to have been completely indifferent to the warnings and pleas of German eugenicists; the numerous calls for eugenics-related reforms, including proposals designed to promote population growth, went unheeded."[62] Eugenics was now invested with a new power to supply remedy against the shock of war and the realisation of national decline; it became an active, dynamic principle assisting the emergence of a racialised political biology.

Indeed, eugenic diagnoses of war proposed after 1916 were made to obviate existing vicissitude, to banish the memory of the degenerated and weak nations, and to systematise the bio-political power of the state. As Neil MacMaster observed, during the war, "[t]he state assumed ever-increasing powers to intervene within the private sphere of the family and maximize reproductive powers through a range of interventions, from compulsory schooling, provision of school meals and milk, family allowances and maternity leave, to restriction on female and child labour, training of midwifes and food hygiene legislation."[63] Eugenicists shared this vision of the state as the laboratory for the nation's rehabilitation, assuming that only the state could mobilise and manage the vast restructuring of society needed during and, especially, after the war.

It was largely due to the demographic changes brought about by the war that political elites turned to eugenics as a means of promoting social and biological revivalism amidst a disillusioned political environment. Social reformers and eugenicists, in turn, alerted the government to the need for a rigorous health policy integrating hygienic and eugenic principles. What ensued was an intense debate not only about national health and societal protection but, ultimately, about national survival. For these measures to be adopted, eugenics had to be institutionalised. The architects of this eugenic process emphasised that society's purification with a view to its biological continuity depended upon the transmission of new racial codes and morality to the general public.

To illustrate this eugenic activism, it may be worth discussing some of most relevant eugenic theories on healthcare for mothers and infants in greater detail. An increase in births was one of the goals of eugenic welfare measures that were to assist those who wanted to have more children. Yet was the state's insistence on childbearing an appropriate eugenic measure? For feminist eugenicists, like Helene Stöcker, racial health was conditioned not only by selective breeding but also by the emancipation of women and the dissemination of a new sexual morality.[64] Although criticising the state for abusing its role as the protectors of women and children, this new cult of motherhood dovetailed with the eugenicists' promotion of racial improvement and a sense of responsibility towards future generations.

Considering how other eugenicists saw women as sources of racial rejuvenation can help further to clarify this point. In 1915 the prominent Hungarian gynaecologist János Bársony contributed a study entitled "Eugenics after the War" to the *Journal of Women Studies and Eugenics*, in which he claimed that the war had destroyed the "healthy and strong men of the nation." Racial fears were thus seemingly justified by statistical evidence about the increase in number of the "inferior individuals" in the population, and eugenics, Bársony opined, needed to respond efficiently to wartime challenges and traumas. Two techniques were drafted to ensure the "recovery of the race." The first course of action was to increase the birth rate. "In Hungary for example, the family with six children is regarded as a rarity in contrast to the past, and there are entire regions in which the 'one-child system' dominates."[65] Additionally, some of the factors contributing to "the stagnation of the Magyar race," such as "birth-prevention, abortion and abortionists," were to be neutralised by preventive eugenic measures. The second approach underlined precisely this point: "The new generation should be not only large, numerically speaking, but also primarily healthy. The health of the parents is the first condition for [the recovery of the race] to happen."[66] More generally, the reappraisal of the mother's eugenic role resulted in a nuanced evaluation of the relationship between eugenics and maternity. There was thus a convergence of interest between the nation's future and the protection of the mother. In order to raise the racial quality of future generations, Bársony advised the Hungarian government to "begin by protecting women."[67]

In Italy, the project of protecting women, as with all other eugenic schemes, involved mobilising the medical profession and state resources. This is particularly well illustrated by the Italian journal *Modern Gynaecology*'s special 1917 issue on the "protection of women and race." Its editor Luigi Maria Bossi problematised the protection of women in terms of its social and legal consequences:

> The defense, therefore, of women and race, in relation with neo-Malthusianism, criminal abortion and the right to abortion of women systematically violated by the Germans constitutes a great,

complex problem that must be resolved through three indivisible relationships: social, juridical and medical. And it is above all pertinent to gynaecologists, because they are responsible, as is obvious, for the basal concept of conservation of the species, of the present life and health, of the mother, and subordinately, of the life and health that is the product of conception. The social and juridical sides must naturally be subordinate to the gynaecological side.[68]

The conflict between the role of women as racial guardians of the nation and physical abuse created by invading armies became particularly evident in the eugenic propaganda directed at the future generations. Unborn children not only served as the ubiquitous emblem for the nation's regeneration, but provided eugenicists with a mobilising medical agenda. As Wilhelm Schallmayer insisted, the existing political elite's priority should be to use eugenic propaganda to create a sense of social responsibility towards future generations.[69]

Some of these practical activities were reflected in the social agendas pursued by various organisations dedicated to the health of mothers and infants, including the German League for the Protection of Mothers (established in 1905); the Hungarian League for the Protection of Children (established in 1906), the Czech Provincial Commission for Child Welfare (established in 1908), the British Association of Infant Welfare and Maternity Centres (established in 1911), and the Turkish Child Protection Society (established in 1917). Building on the reproductive role of women and various strategies to decrease infant mortality, many of these associations, like the eugenic organisations, provided a comprehensive catalogue of nationalist virtues.[70] The devotion of motherhood to the patriotic cause contributed not only to the strengthening of nationalism but also enabled states to control sexual reproduction and proposed a normative ideal of femininity.[71] But, as Kristen Stromberg Childers argued in the case of France, "[w]omen were not the only 'gendered' beings" in interwar eugenic, natalist and hygienic discourses. Beginning with the First World War, "French legislators made every effort to cast men in certain gender-specific roles that would enhance the protection of the family and, ultimately, the nation."[72] As that biopolitical unit, the large family, was being carved out of the social aggregate, eugenics reinstated the national extendedness of family life.

Towards the end of the war, the eugenic language, initially the privileged property of physicians and biologists, was increasingly being adopted by various authorities of social and cultural life as well as by national propagandists. As the hereditarian vocabulary increasingly shaped debates on fertility, children and their social protection, it also underpinned new nationalist discourses that equated large families with a strong nation, and compared the one-child system with outdated cultural traditions. Given such concerns, medical experts viewed the protection of future generations both in social and national terms. It justified not only immediate measures against birth control, but also fed eugenic ideas of national rebirth. Under the pressure of war, eugenicists believed that the nation's demographic potential would be seriously undermined unless the state provided coherent policies towards the protection of mothers and infants.

The birth of healthy children was increasingly viewed less as an exclusively private matter, and more as a matter of major concern to the state. One such example is the introduction of a National Baby Week in 1917 in England as a result of the efforts of the Eugenics Education Society. This was an opportunity for eugenicists and other health reformers to organise "meetings, exhibits and lectures on prenatal and postnatal care" as well as direct attention to "the need for healthy children to fill the stilled cradles of the land."[73] Moreover, this general public's growing interest in eugenics was reflected by the abundance of eugenic organisations established during the war, as well as the numerous national and international conferences on health, hygiene and population policies.[74] In 1917, the Hungarian Eugenics Society was transformed into the Hungarian Society for Racial Hygiene and Population Policy.[75] In several respects, this eugenic society differed from similar organisations in Europe. As Hoffmann proudly announced, "the double movement which divided the efforts of race regeneration in Germany was united in Hungary from the beginning."[76] According to Hoffmann, a conceptual and practical delineation between eugenics and populations policies, as was the case in Germany and Britain, was detrimental to the evolution of the Hungarian eugenic movement. Reflecting this symbiosis, the Hungarian Society for Racial Hygiene and Population Policy had three main objectives. It campaigned for:

1. The scientific exploration of those damages that threaten the body of the Hungarian nation, particularly the declining birth rate;

2. Establishing the means and ways by which to increase the number of births; 3. The support of those endeavours whose purpose was the creation of an environment in which the Magyar race could prosper.

Ultimately though, what counted was "that race-consciousness, the consideration for future generations and the high estimation of proficient big families, was to be inculcated into all branches of the state, social, economic, political and moral life."[77] That this transformation of the general public's perception of eugenics was not effortless was further accentuated by István Apáthy: "Eugenic measures often require a strong heart and hard faith in human evolution, and this is nothing but the self-abnegation, which the pusillanimous and the masses are always inclined to condemn as callosity."[78] This pragmatic eugenic utopianism was based on firm moral principles but equally conscious of the obstacles which lay ahead on the road to practical eugenics.

But governmental assistance was essential to eugenic work. In Hungary, the Society for Racial Hygiene and Population Policy received most of the institutional support it needed to spread eugenic propaganda from the Ministry of War. So, for instance, and in response to the serious problems affecting Hungarian civilians during the war such as contagious diseases and mortality, the Ministry created special commissions to promote the well-being of the family, including marriage counselling and medical assistance for venereal infections.[79] These objectives were pursued, to different degrees, by most European nations, indicating how rapidly eugenic ideas had spread towards the end of the war.[80] New organisations joined established ones, such as the Polish Society for the Struggle against Race Degeneration founded in 1917, expressing not only the struggle to build a new nation, politically and institutionally, but also morally and spiritually.

In many ways, and as Modris Eksteins wonderfully demonstrated, many European nations had considered war a "spiritual necessity;" but for the eugenicists who marshalled to revitalise the nation it was also a "rite of spring." Modris's inspired analysis of the modern culture germinating during the First World War as the "exuberant eruption of life that comes with the awakening of spring"[81] is also extremely useful in understanding the complementary relationship between modernism and eugenics. For where modernism valorised war as a source of cultural rejuvenation, eugenics aligned its dynamism with the moral and biological transformation of the nation.

The Dawn of a New Era

The numerous wartime debates on the theoretical or practical merits of eugenics are revealing. On the one hand they produced a diversity of interpretations of eugenics and its immediate social and political purpose that illustrate the importance the racial sciences had acquired in national politics in various European countries between 1914 and 1918. On the other hand, and due to specific wartime circumstances that emphasised the importance of national survival, these debates help explain how eugenics became politicised. The war not only anchored eugenics in social and political fields but, and equally important, created an auspicious environment for the idea of the resurrected nation to take hold.

As one of the nations most severely affected by war, both territorially and demographically, Hungary was certainly an exemplary case. At the Third International Congress of Eugenics held in 1932, the medical statistician Tivadar Szél argued that the eugenic effects of war mostly affected the "members of the Reformed Church", of pure Hungarian racial origin, as these were the individuals "burning with patriotic zeal" and not "the Jews who avoided the war by means of exception."[82] Not surprisingly then, Szél believed that

> the Great War caused the re-awakening of the nation and developed its race recognition, and that after the great afflictions the central power is rising with renewed vigour to a consciousness stronger in many respects than before, and taking measures to protect the nation and the race as well as to improve the quality of future generations.[83]

Accordingly, eugenics not only claimed to represent the racial effort of the wartime generation, it also embodied the spirit of the new nationalism intrinsic to the resurrection of Hungary.

Szél explicitly claimed that the spiritual qualities forged by experience of war, albeit dramatic, supposedly endowed new generations of Hungarians with the dynamism and willingness to pursue the radical transformation of their national consciousness. "One thing," he concluded,

> is certain: the numerical damage wrought to the population by the war losses, plus the territorial losses by virtue of the Peace Treaty, were more severe for Hungary than for any other country. Notwithstanding this tremendous depopulation, the Magyars survive, with an unshakeable

faith in a better future, trusting in the eventual revision of the unjust Peace Treaty. Never was the Magyars' position today, after their unprecedented war losses, better characterised than in the words of the great Hungarian poet Michael Vörösmarthy: Diminished yet unbroken/Lives the nation in this land.[84]

For Szél – lamenting the predicaments of a defeated nation – the war embodied a new era that augured the emergence of a new revolutionary community of Hungarian nationalists. This was a detail succinctly captured by George L. Mosse as well when he noted that "[t]he memory of war was refashioned into a sacred experience which provided the nation with a new depth of religious feeling, putting at its disposal ever-present saints and martyrs, places of worship, and a heritage to emulate."[85]

Comparing the various eugenic interpretations of war discussed in this chapter, we do not find a clear consensus or linear progression, but a myriad of tensions, disagreements and, above all, uncertainty. But this diversity and complexity were also sources of strength, and they suggest that eugenics was emerging as a powerful ideology in war-troubled Europe through its capacity to hold together, in a potent synthesis, a range of social, cultural and political concepts whose coexistence, at first sight, may appear impossible. After the First World War, Europe faced a series of destabilising moments, including revolutions in Central and Eastern Europe, the collapse of multinational empires, civil wars and the emergence of new nation states. As we shall see in the next chapter, subsequent historical developments during the interwar period, most notably territorial disintegration, population losses, political revisionism, racial legislation and anti-Semitism, proved, however, that the call for race regeneration voiced by many during the war was more than mere eugenic rhetoric; it became an essential component of national politics in many European states.

3

EUGENIC TECHNOLOGIES OF NATIONAL IMPROVEMENT, 1918–1933

This chapter concentrates on the practical application of eugenics and its relationship to the idea of national regeneration, with a particular focus on the period between 1918 and 1933. According to Emilio Gentile, "[d]uring the twentieth century, national regeneration was presented as a total revolution to be achieved by one of two means: by new culture or by new politics. Until the Great War, the first method prevailed. After the war, with fascism, the new politics – totalitarian politics – claimed for itself the task of regenerating the nation."[1] There was, however, another method of national and racial improvement, which Gentile does not discuss, namely science. It does not suffice to explain modernist nationalism through the themes of rejuvenation created by avant-garde poets, artists or novelists. As I have argued elsewhere, eugenics aimed to create a national ontology wherein the nation as an object was paramount. By offering a physical representation of the nation eugenicists engaged in allegedly objective incursions into the ethnic fabric of society, contrasting their diagnoses of modernity's troubles with those offered by literary texts or artistic images.[2]

After the First World War, eugenics intensified its regenerative content, verbalising its ambition to reconfigure the national community less with notions of egalitarian participation in public life than with programmes based on the biological selection of valuable racial elements.[3] In the name of science, eugenicists synthesised hereditarian determinism with the modernist political revolution, insisting that both pursued the same goal: to seal the societal and cultural chasms torn open by modernity. Whether in the form of national celebrations, as in the case of the victorious nations, or collective mourning, as in the case of the defeated countries, eugenicists heralded the reconciliation between state and nation, positing science as a solution to a number of problems nations faced after the war.[4] Bemoaning that the nation's perceived decline was intensifying, eugenicists called for immediate legislation to prevent the social collective's further deterioration. They campaigned vigorously for the nation's social and biological improvement and, as we shall see, in many instances they were successful.

Eugenic Stigma

At the Fifteenth International Congress of Medicine gathered in Lisbon in 1906, Ladislav Haškovec, the most prominent and influential Czech eugenicist of his day, presented his views on marriage restrictions which he had already published in a Czech article in 1902. In Lisbon, he meticulously demonstrated the relationship between heredity and disease, insisting on the importance of restructuring social hygiene and public health around the developing ideas of eugenics. "We must go down to the root of evil," Haškovec summoned his colleagues. "If," he added, "in founding the family, consideration is given to the consequences of pathological heredity and to congenital maladies, one may absolutely affirm the diminution of the number of feeble-minded persons, of syphilitics, of tubercular persons, of criminals, and of children afflicted with nervous diseases or otherwise degenerate."[5]

As eugenics' popularity grew amongst European physicians and intellectuals after 1900 so too, in similar fashion, ideas of normative health and racially perfect communities matured amongst eugenicists. With the growing acceptance of heredity as the main factor in determining any individual's physical and psychological evolution, Haškovec and others could turn to the social and biological project of improving the population according to the Mendelian laws of inheritance. The Swiss psychiatrist Auguste Forel, for instance, was renowned for his elaborate eugenic edifice designed to protect society from the danger of deviancy and biological malfunction. In his 1905 *The Sexual Question*, Forel entrusted the state with medico-legal powers over criminals and asocial individuals, resting on the assumption that the most efficient method to regulate the distinctions between "normality" and "abnormality" is the drastic castigation of the latter. As a pioneer of sterilisation and castration proposals, Forel's theories of reproductive sexuality and biological protectionism were built on substantively eugenic foundations. Arguing against those who might have criticised eugenics' anthropological project, Forel reiterated that "it is not our object to create a new human race of superior beings, but simply to cause gradual elimination of the unfit, by suppressing the causes of blastophthoria, and sterilising those who have hereditary taints by means of a voluntary act; at the same time urging healthier, happier and more social men to multiply more and more."[6]

This vision of human perfectibility was central to representations of the national body as incessantly under threat from various forms of pathology and innumerable diseases. Relating these to themes of heredity and degeneration, and to ideals of health and hygiene, Forel described sterilisation as the regulatory mechanism to ensure racial survival. The extent of his medical scrutiny meant that he recommended the sterilisation of various "categories" ranging from "criminals, lunatics and imbeciles" to "individuals who are irresponsible, mischievous, quarrelsome or amoral," or even "persons incapable of procreating a healthy race owing to inherited diseases or bad constitution."[7]

Having established a biological taxonomy dividing society into two camps, health and physical beauty became the standards governing the nation's racial welfare. The othering of those deemed biologically "unfit" and "degenerate" was an essential component of a campaign of eugenic stigmatisation driven by the state and eugenicists to draw a new, biological map of the body of the nation. Eugenic stigma highlights a flawed social and biological identity, an identity that was marked in a specific cultural and national setting. Those thus eugenically stigmatised were excluded from the rights and privileges of the nation. In his classic 1963 study on *Stigma* the sociologist Ervin Goffman identified three different types of stigma:

> First there are abominations of the body – the various physical deformities. Next there are blemishes of individual character perceived as weak will, domineering or unnatural passions, treacherous and rigid beliefs, and dishonesty, these being inferred from a known record of,

for example, mental disorder, imprisonment, addiction, alcoholism, homosexuality, unemployment, suicidal attempts, and radical political behaviour. Finally, there are the tribal stigma of race, nation, and religion, these being stigma that can be transmitted through lineages and equally contaminate all members of a family.[8]

One easily recognises the complete repertoire of medical and social identifiers constituting the object of negative eugenic discourses here. As early as 1873, Galton advised those belonging to "inferior races" to maintain celibacy, but if they "continued to procreate children, inferior in moral, intellectual and physical qualities, it is easy to believe that time may come when such persons would be considered as enemies of the State, and to have forfeited all claims to kindness."[9] But it is important to realise that eugenic stigmas were not just the result of a medical terminology invading the social and political spheres, but also the outcome of an essentially modern process that I termed the biologisation of national belonging. In order to protect the nation from those deemed "unhealthy," "diseased" or "anti-social," eugenicists constructed "a stigma-theory, an ideology to explain [...] inferiority and account for the danger [it] represents."[10] The external attributes of physical or mental infirmity were accentuated in order to legitimise eugenic action against individuals who did not conform to the normality of the national community. The individual who was eugenically stigmatised was an individual whose biological and social identity was called into question and castigated accordingly. Take, for example, the following descriptions of the mentally disabled collected by a Finnish committee on sterilisation appointed by the government in 1926. One woman was marked out as an "imbecile, twenty-seven-year-old daughter of a butcher; peaceful, could be nursed in freedom at home were it not for her eagerness for sexual intercourse; therefore recently she had been cared for in a district lunatic asylum; at home, gave birth to an illegitimate child." Another person was described as a "middle-aged insane man, who believed he would be cured of his illness if he were allowed to have sexual intercourse with a virgin or a pregnant woman."[11]

But eugenic stigma was unsystematic. Social and biological degeneration, and the danger it posed to the national community was construed differently in each national context. Hygienic strategies devised to secure national progress in Finland were often deemed inadequate in Romania. But however different such strategies were, a universal component united them, namely the eugenic power invested in the state. Eugenics and the state were mutually maintaining agents in that eugenic reforms required both a proper programme of social engineering and the means to implement them. The Austrian eugenicist Ignaz Kaup explicitly connected welfare institutions with racial protectionism when he criticised a political economy that neglected the costs "inferior elements" burdened state and society with. "Our healthy offspring have a right to protection from corruption from the genetically damaged, and every progressive nation has the duty to avoid the burden of the costs of inferiors as far as possible."[12] Echoing Giuseppe Sergi's call for state intervention voiced during the war, Kaup hoped that the state would finally mobilise and actively engage with the national transformation. The state, he argued, should be transformed not only in terms of its infrastructure, economic development and political institutions, but also in terms of education, public health and modern hygiene imperatives.

Eugenics, as one of the most efficient strategies of controlling the biological purpose of the individual, pointed to a future homogeneous society in which the nation's racial health became the dominant principle of government. The 1913 Marriage Bill proposed by the Swedish Marriage Commission is illustrative here:

> With modern racial hygiene, eugenics, legislative precautions are urgently recommended to protect future generations, and to preserve and improve the human race. This movement wants to fight not only such threats to the public health as emigration, industrialism or the accumulation of people in urban centres but also racial poisons as syphilis, tuberculosis and alcohol. Thus it wants to encourage society to consciously work at increasing the marriage frequency among its better citizens as it wants to prevent the propagation of the unfit.[13]

This political biology articulated new social and racial genealogies of identity indicating, on the one hand, how eugenic discourses of the late nineteenth and early twentieth century were transformed after the First World War into a totalising national narrative about biologically redesigned societies and, on the other, how this narrative legitimised the projects of national regeneration to be implemented on behalf of a new racial community.

Post-War Reconstruction

Paralleling this process of eugenic stigmatisation was another one: the process of national reconstruction after the war. "National and social renaissance," observed the Croat health reformer, Andrija Štampar, in 1919, "is at the same time health renaissance."[14] The health of the population became the central component of new national welfare programmes devised during the interwar period. Yet this concerted focus on health not only transformed the significance of population as a site of biological power but also systematised an approach to eugenics that relied on state intervention in the private and public sphere. Gradually, eugenics peeled away the layers of culture to produce a political biology based on nature.

Moreover, eugenicists reinstated the sacred nationalist connection between identity and territory. They, like other professionals, were deeply affected by the post-war rearrangement of Europe's political landscape.[15] Some rejoiced; others rebuked. In 1921, the German geneticist Erwin Baur wrote a rejoinder to Harry Laughlin's overview of recent German trends in eugenics published in *The Eugenics Review*.[16] It was a sombre reflection on Germany's situation immediately after the war and the evolution of eugenics in that country.[17] Lamenting the foreign occupation of German territories and precarious economic conditions, Baur nevertheless found a number of reasons why these calamitous circumstances did facilitate an increase of eugenic awareness amongst Germans. "The very magnitude" of Germany's military humiliation, Baur believed, contributed to "a greater understanding of eugenic questions than before the war." Another factor was that "impoverishment" and "severe distress" that had "acted favourably in a eugenic sense." The terrible famine that resulted from "the English blockade" in the winter of 1916–1917 was depicted as "an abomination [Scheusslichkeit], yet it probably had a favourable influence" eugenically. Finally, Baur envisioned an improvement of national morality assuming that pre-war "contemptible luxury, the nauseating search for pleasure, the modes of thought and ways of life directed solely towards superficialities and coarse sensualities [Grobsinnliche] will probably dwindle on account of the poverty of the people."[18]

Heralds of new cultural and political landscapes emerging from the war, eugenicists – like other modernists – reaffirmed their commitment to a vision of society that was racially healthy and morally powerful, one able to withstand what Marshall Berman described as modernity's "maelstrom of perpetual disintegration and renewal, of struggle and contradiction, of ambiguity and anguish."[19] Although Baur's glossing over the modernist theme "all that is solid melts into air" exemplifies the war's profound impact on most Germans, his view is also informed by a revivalist eugenic episteme. It is not going too far to suggest that in addition to anticipating the widespread acceptance of eugenics after 1918 in Germany and elsewhere, Baur also visualised a new eugenic praxis, one exclusively geared towards protecting one's nation and race. The Czech eugenicist Vladislav Růžička concurred: "As the world is undergoing a regeneration and everything is being organized and reorganized, it is imperative to also improve the organization of Eugenics. This demand is especially justifiable as eugenic research is not only concerned with the evils caused by the war, but also about the questions connected in general with the substance of national and racial being."[20]

To some, this national rejuvenation was to be achieved by creating a system of public health that detected the prevalent social illnesses and acted upon adequate racial laws to alleviate them. The German eugenicist Hermann Siemens, for one, insisted that "[r]ace hygiene must vigorously demand that 'punishment' shall cease to be the purpose of criminal law. We much rather require an administration of justice that has for its purpose the protection of the race. To render pathological natures permanently harmless and to prevent the reproduction of other miserable creatures should consciously become to goal of our courts."[21] Those afflicted with hereditary illnesses were considered purveyors of various social and medical maladies.

Although several governments proved reluctant to enunciate negative eugenic guidelines for post-war reconstruction, eugenicists were not discouraged. Whilst discussing Hungarian eugenics with an American audience in 1920, the veteran supporter of sterilisation Géza von Hoffmann[22] lamented that too "[m]uch stress is laid upon the positive side of the question, i.e. the propagation of the fit, and no steps have yet been taken to cut off the propagation of the unfit."[23] But the American biologist Raymond Pearl, commenting on sterilisation's genetic utility in the first post-war issue of *The Eugenics Review*, prudently cautioned that "whilst compulsory sterilisation is the only adequate means yet suggested for the prevention of the reproduction of the socially unfit, striking immediate results in the reduction of the number of degenerates and defectives are not to be expected to follow the inauguration of such a system. Many years must elapse before the proportion of hereditary degeneracy in the race could be substantially reduced."[24]

Without an international conceptual consensus to inform it, both controversial and often condemned on moral and religious grounds, negative eugenic theories of national improvement were rarely discussed during the First World War. But during the economic crises and political instabilities characterising the 1920s, eugenic sterilisation began to attract considerable attention from both the medical profession and social reformers interested in protecting the nation from an alleged biological degeneration and social decline.[25] Albeit contentious, negative eugenic theories were gradually granted a central role in the modernist drama of national regeneration unfolding in most of Europe during the interwar period.

In Germany, according to Paul Weindling, defeat in the First World War "gave eugenics relevance with regards to national reconstruction. Virtually every aspect of eugenic thought and practice – from 'euthanasia' of the unfit and compulsory sterilisation to positive welfare – was developed during the turmoil of the crucial years between 1918 and 1924."[26] In Italy, post-war social and biological regeneration agendas coincided with the struggles to combat social chaos and political disintegration. As the participants at the first conference on social eugenics in 1924 indicated, the state must be given new prerogatives if it were to ensure national reconstruction:

> The experience of the war and the needs of reconstruction have put in a new light the importance for a State of the physical and psychic qualities of its citizens and have therefore called the attention of men of government and scientists to studies and measures directed to improve in a permanent way the health conditions, labour efficiency, and the gifts of intelligence and will of the population.[27]

Many eugenic justifications for social and biological engineering were then taken over by eugenicists and government officials alike. Themes of economic sustainability were coupled with new visions of society, ones that reflected the anticipation or desolation many Europeans felt after the war. In Germany, for example, the campaign for sterilisation took a new, more vigorous turn by the end of the First World War. It was then that "German aggrandizement and stability seemed at its lowest," and that "sterilization was widely and passionately recommended as a solution to urgent social problems."[28] By the late 1920s voluntary sterilisation "had become the ideological leading edge of the eugenics movement" in Britain as well. Moreover, as John Macnicol noted, once scientific evidence established the hereditarian nature of the mental defective, and their voluntary sterilisation was pursued, "further negative eugenic measures could be implemented – ultimately (though there was much disagreement on this), compulsory sterilization for other categories of the unfit."[29] Quintessentially, supporters of practical eugenic measures, like matrimonial health certificates, segregation and sterilisation, maintained that they were rendering the utmost service to society: by defending future generations from social and biological degeneration.

Whether such authors thought in terms of positive or negative eugenic policies, there was general consensus that collective eugenic action to transform the nation was necessary. Indeed, as the widespread support for technologies of social and biological improvement grew during the interwar period, eugenic sterilisation was as enthusiastically supported by nationalists in Romania as it was by social democrats in Sweden, or by religious leaders and atheists in Finland. It should not, however, be assumed that this consensus on the practical applicability of eugenics obliterated the emergence of particular national eugenic agendas. Both in theory and practice the increased nationalisation of eugenics after 1918 points to racial fears and national anxieties specific to the nature of various European societies. Post-war national reconstruction efforts after the war inspired a series of eugenic explorations of possible national futures in most European countries, but these visions often competed with each other, both domestically and internationally.

The Nationalisation of Eugenics

As discussed in the previous chapter, although the nationalisation of eugenics intensified during the war, it was only with the political and territorial transformations introduced by the peace treaties of 1919–1920 that eugenicists changed strategy by turning their attention to political agitation and aggressive nationalism. International credibility was certainly very important to eugenics, but it was not an end in itself. Above and beyond the universal recognition of their respective scholarly fields, eugenicists – like most scientists – were driven by a genuine commitment to improve the health conditions of their own countries and nations. The key to this practical form of national medicine was health reform and hygiene education, combined with the necessity of establishing a system of eugenic schooling and research that would provide the population with the knowledge of the scientific principles that governed their physical and social existences.

Shortly after the establishment of the Czechoslovak state on 28 October 1918, the Czech Eugenics Society petitioned the Republic's president, Tomáš Garrigue Masaryk, with the following requests:

> a) the creation of a national institute of eugenic research; b) the adoption of special records for registering the health of the population; c) the adoption of central eugenic stations; d) the creation of institutes for the study of the development of human psychology, as well as a museum of comparative genetics; e) the protection of infants; f) the reform of midwifery; g) the reorganisation of the system of teaching modern hygiene, especially in terms of sex education; support for eugenic instruction in society by means of public discussions, theatrical and cinematographic performances, and, in particular, the establishment of a Museum of Hygiene as the centre-point of all instruction; and, finally, i) the compulsory issuing of a health certificate before marriage.[30]

Within this extensive eugenic programme, the nation became the object not only of a rational pedagogy, but of a spatiotemporal transformation, one that reflected the practical realities of post-war Europe. Eugenicists had no more pressing problem than awakening the nation to the practical necessity of a biological rejuvenation built around the laws of heredity. Germany, again, is a resourceful example, as it is in this country that interwar eugenic attempts to completely transform the role of the individual and the family within the nation's overall biological body enjoyed their greatest successes.[31]

The German League for National Regeneration and Heredity, established in 1925, eloquently illustrates how eugenic ideals were incorporated into post-war discourses on degeneration and its detrimental effects on the nation and future generations. The slogan "Protect German Heredity and thus the German Type" symbolised both eugenic scientism and the aspiration for a racially homogeneous national community. In addition to eugenic education, the League supported demands that those deemed physically and mentally degenerate should not be allowed to reproduce, and that regulating marriage practices could prevent those with hereditary diseases from ruining future families. A similar association was established in Austria in 1928 that, like its German counterpart, advocated educating the masses on the ills of social and biological degeneration, and the preservation and improvement of national racial qualities.[32]

Although most eugenics societies survived the war, including the Eugenics Education Society and the German Society for Racial Hygiene, others, like the Hungarian Society for Racial Hygiene and Population Policy, were left dismembered. Numerous others were founded, including the Italian Society for Genetics and Eugenics in 1919, the Swedish Institute for Race Biology in 1921, the Czechoslovak Institute of National Eugenics in 1923; both the Viennese Society for Racial Hygiene and the Estonian Eugenics Society were formed in 1924. In Poland, a Section for Social Hygiene and Eugenics was launched by the Society of the Health Protection of the Jewish Community in 1918, whilst the Society for Combating Venereal Diseases was renamed the Polish Eugenics Society in 1922. In Romania, the Eugenics and Biopolitical Section of the "Astra" Association was established in 1927, followed by the Bulgarian Society for Racial Hygiene in 1928, and the Belgian Society for Preventive Medicine and Eugenics in 1929.

As European states were slowly recovering from a devastating war, it became apparent that a more dynamic eugenic strategy was needed to meet current needs. It was also evident that counteracting the increased number of physically and mentally incapable, criminals, paupers and orphans was imperative. And, perhaps most importantly, the state had to take preventive measures to stem what many eugenicists saw as a frightening spread of degeneracy. To illustrate this new eugenic activism, it is worth quoting from the motion presented to the King of Sweden by the Swedish Parliament, endorsing the creation of the Institute for Race Biology in Uppsala:

> Race-biological investigation, which works to attain a high and noble object: protection against genealogical degeneration, and the furtherance of good racial attributes, grows daily stronger and stronger. The motion is based upon the standpoint that there is nothing so precious in a country as the racial material itself, especially if this, as is the case with

the Swedes from ancient times, is of good quality. The task this scientific investigation has to contend with is to explicate and elucidate all conditions regarding heredity and environment which cause an elevation, or debasement of the inherent worth of a race. Then, firm bases and guidance can be given to a State in its endeavouring to enhance the development of the race and people in the right way. Race-biology is, in other words, the scientific study of all the factors which exercise influence on the physical and spiritual structure of coming generations.[33]

The organisation of these eugenics societies and institutes, in addition to popularising ideas of health and hygiene, was an integral part of the development of what can be called the nationalisation of eugenics, namely the transformation of eugenics into a "national science" devoted to the protection of the specific nation's health. The eugenic conceptual framework continued to be based on universal principles like humanity, but its aims were now increasingly directed towards transforming the geography of the national space. As Vladislav Růžička put it in 1923, eugenics must "become *national eugenics* in the most literal sense of that word. [Its] aim is to maintain and develop the biological individuality of a nation and prevent the decline of its biological organization in whatever respect."[34]

The Serbian hygienist Vladimir S. Stanojević underlined this argument in his 1920 *Eugenics* when he noted: "The hygienic refinement and improvement of descendants – this is the future religion for the individual and the family as well for the whole of cultured humanity."[35] Through fixing the process of creating a new national being in scientific language, eugenics enacted a new nationalist platform, namely a cultural critique of modernity and its seditious replacement. Eugenic attempts to create a new "Yugoslav race" from the fusion of Croat, Serbian and Slovene nations exemplify such a process, whereby eugenics effected the implications of a broader national enactment of biological metanoia. As one contributor to the Croatian journal *New Europe* explained in 1924: "According to the laws of contemporary eugenics, progress is achieved by the process of mixing differed but closely related tribes to produce a physically superior type. [...] This mixing will be profitable to us. It is apparent, for example, that the bony, stocky and militant Dinaric type will strengthen the average Yugoslav person just as the strongly evolved Slovenian women will."[36]

The increased politicisation of modern technologies of health facilitated the complex mechanism of translating scientific idioms into diverse national contexts. Inaugurating the new journal *Annals of Eugenics* in 1925, Karl Pearson explained that Galton prefixed eugenics with the adjective national because "he conceived that the nation, not the family nor the individual, was the proper unit for study [....] Even if we allow that much eugenic thinking finds its application through the individual and the family, it will still appear that the mainspring of eugenic doctrine has national rather than individual welfare as its motive; it aims at the betterment of future generations rather than at the increased comfort of the individual."[37]

Pearson was not the only eugenicist aspiring to award the nation's biological identity the centre stage of the public's imagination. Different cultural and religious affiliations notwithstanding, virtually every eugenicist in interwar Europe aimed to build a national eugenic paradigm. In his 1925 *The Hygiene of the Nation*, the Romanian eugenicist Iuliu Moldovan conceptualised eugenics in exclusively national terms by connecting it to biopolitical interventionism and radical measures to regulate health.[38] Moldovan placed the family at the centre of his theory of national eugenics, envisioning measures to protect "acceptable" families from both social and biological threats. However, Moldovan claimed that prophylactic medical methods would not suffice unless the Romanian population – especially those affected by hereditary diseases – acquired "a racial consciousness, a sentiment of biological responsibility."[39] Similarly, the German racial anthropologist Otto Reche's opening address to the Viennese Society for Racial Hygiene in 1925 emphasised the state's responsibility to sustain a healthy racial stock, whilst condemning individualism, social deviance and inherited disease.[40] This orientation was also shared by the Hungarian economist Károly Balás, who declared that "[e]very state, nation or society which recognises its own highest interests devotes considerable attention to the question of racial maintenance and racial protection."[41]

Although certainly motivated by national concerns, this interest in one's own national community has to be seen in context with the broader ambition to ground social and institutional policies in the laws of heredity and eugenics. As the Greek eugenicist Stavros Zurukzoglu pointed out in his 1925 *Biological Problems of Racial Hygiene*, modern eugenics should have the two central aims of instilling racial responsibility towards the nation as well as ensuring the harmonious growth of each individual's physical, moral and intellectual capacities. This two-pronged approach was fully consistent

with the modern temporality of eugenics, which Zurukzoglu visualised as an efficient antidote to the social and economic difficulties
of post-war daily life.[42]

By setting out the various eugenic tasks vital to rebuilding post-
war societies, eugenicists like Pearson, Moldovan and Zurukzoglu
demonstrated that the quest for social and biological improvement
as tailored to state prerogatives had to include both extensive eugenic
propaganda and programmes of racial nurturing. Practical eugenics demanded that all social spheres be tied to the medical scrutiny
of the health technocrats supported by official laws. Applied to
national education, eugenics' most important role was to promise
social improvement and inspire public confidence in the country's
future. Outlining his vision of social improvement in 1926, the
German social hygienist Alfred Grotjahn, therefore, placed "practical eugenics" at the intersection of social hygiene and social politics,
connecting it to "the current questions about our country's national
existence and prestige."[43]

Such a wide national project of social engineering demanded a
corresponding biological crusade against alleged degenerations of
the national body. The nationalisation of eugenics occurring during the 1920s was as a result congruent with an intensification of the
biologisation of national belonging. But in its quest for the perfect
nation eugenics also promoted more extreme methods of biological
cleansing. In this climate, then, radical eugenicists reiterated their
desire to refashion the body politic, and to thereby drive the nation's
rebirth. The Swedish eugenicist Herman Lundborg expressed this
clearly when he stated that:

> The carrying-out of race biological research should be a State duty.
> The desire for self-preservation will sooner or later lead the people of
> all civilized countries to establish institutes of race biology. We may
> then arrive at the firm conviction that the nations which early enough
> realize the importance of inherited health and act upon this realization
> will advance triumphantly; they will attain to a higher morality, a more
> intelligent culture, a healthier and happier state of being.[44]

If eugenics was the art of achieving human improvement, and
accordingly involved the nation's biological transformation, it was
obvious to eugenicists that the state's ultimate supremacy in dictating, controlling and implementing this transformation was unques-
tionable. The eugenicists' social role changed considerably, and

correspondingly they became experts of the "socio-biological sciences," which in turn decided which parts of the population were "racially" valuable and which ones needed correction. As Lundborg remarked as early as 1904, "I dare hope that the time is not far distant, when one will be inclined, in public affairs, to allow the word of the biologically educated physician to have as much weight at least as that of the lawyer and soldier, and when sociologists and statesmen awake to the significance of heredity-hygiene for the future of mankind."[45] By the late 1920s, the physician had become not only an esteemed social actor, but the key instrument of state-sponsored biological engineering. As the Romanian health reformer Iuliu Haţieganu put it in 1925: "[t]hrough his career, a doctor is the most useful and important social agent of the state."[46] Physicians, Haţieganu continued, would only be able to implement their eugenic ideas when "governments will understand that no progress and no prosperity are possible without seriously organising the state's hygiene and fighting against social diseases [in order to] favour creating a more robust human species and protect the race."[47]

Ultimately, the state's instrumentalisation of eugenics produced a new type of ideologue, namely the biological expert who wished to not only interfere in the life of the individual, but to shape the nation's physical *Weltanschauung*. Nowhere is this ambition reflected more accurately than in Eugen Fischer's memorandum to Benito Mussolini in 1929, occasioned by the meeting of the International Federation of Eugenic Organisations in Rome. "It seems natural and desirable," Fischer assumed, "when considering eugenic problems, that some expression of our hopes and wishes should be addressed to the great statesman who, in the Eternal City, shows more than any other leader today, both in deed and word, how much he has taken the eugenic problems of his people to heart."[48]

This moment was a turning point not only in the history of eugenics and the state, but also of modernism and eugenics. The fascist revolution and creation of the "new man" envisioned by Mussolini dovetailed perfectly with the project of anthropological regeneration advocated by the eugenicists. Fascist and eugenic aesthetics were congruent, as both were centred on the ideal of a healthy, beautiful body. As Pierre Drieu La Rochelle enthusiastically commented in 1941, "the revolution of the body, the restoration of the body" was the "greatest revolution of the twentieth century."[49] The significance La Rochelle ascribed to this regenerated body within fascist aesthetics, however, accentuates another point repeatedly stressed in this study: the normative purpose of eugenics was to not only elate the prospects of producing healthy and beautiful bodies, but offer the means by which these bodies were to be perfected.

Unworthy Life

Few authors of the period produced a more radical interpretation of social improvement than the German jurist Karl Binding and the psychiatrist Alfred Hoche in their ill-famed 1920 *Permitting the Destruction of Unworthy Life*.[50] Offering a legal and economic explanation why euthanasia was to be preferred to other forms of artificial selection, Karl Binding also posed the following question: "[a]re there human lives which have so completely lost the attribute of legal status that that their continuation has permanently lost all value, both for the bearer of that life and for society?"[51] He subsequently identified two main categories of individuals whose legal status and value for society rendered them not merely insignificant, but an economic burden. The first group consisted of "those irretrievably lost as a result of illness and injury," whilst the second included "incurable idiots, no matter whether they are so congenitally or have (like paralytics) become so in the final stages of suffering."[52]

Euthanasia was the extreme expression of the myth of biological regeneration. Binding and Hoche, like others who shared their precepts, expressed their commitment to a new society purged of pathologies and various maladies by requesting the enthronement of new moral principles. "In times of higher morality – in our times all heroism has been lost – these poor souls would surely have been freed from themselves officially. But who today, in our enervated age, compels himself to acknowledge this necessity, and hence its justification?"[53] Further accentuating the hope for a new biological worldview, Hoche concluded that "[a] new age will arrive – operating with a higher morality and with great sacrifice – which will actually give up the requirements of an exaggerated humanism and overvaluation of mere existence."[54]

Admittedly, Binding and Hoche's narratives on economic and social improvement through euthanasia and biological purges were excessive, but certainly not unique amongst European intellectuals of the 1920s. Importantly, moreover, not all schemes of biological renewal were designed to favour one particular race over others. Eugenics has also inspired another vision of biological rejuvenation than one based on the elimination of those deemed "inferior" and "unfit."[55] Though formed culturally and intellectually within the same European environment, this vision was transformed by the belief that sexual harmony and a healthy lifestyle were the founding principles of a new eugenic intimacy. The Transylvanian-born Edmond Székely, who in 1928 co-founded (together with Romain Rolland) the International Biogenic Society, is an exemplary case. Székely's vegetarianism and his plea for metaphysical renewal through meditation and medicine centred on the individual not on the nation or the state. His idea of regenerating a fundamentally corrupt humanity, although eugenic, was directed towards universal not discriminatory spiritual improvement.[56]

It bears repetition that eugenics occasioned a new understanding of humanity that was reflected in a wide range of philosophical, sociological and cultural speculations, often opposed ideologically. Take, for example, the French anarchist and neo-Malthusian Manuel Devaldès. "Sub-humans," Devaldès believed, "were the excretions of society."[57] His Romanian disciple Eugen Relgis similarly maintained that "[t]hese days, with the help of science, degenerates could be exterminated through *euthanasia*. It is, however, preferable, from all points of view, that degenerates should not be born, or, even better, not conceived. And, this is possible with the help of science: by *sterilizing* those who exhibit pathological characteristics or incurable diseases."[58] Finally, the president of the New York Zoological Society, Madison Grant, voiced the concerns of many Europeans when he claimed in his oft-quoted 1916 *The Passing of the Great Race* that: "[m]istaken regard for what are believed to be divine laws and a sentimental belief in the sanctity of human life tend to prevent both the elimination of defective infants and the sterilisation of such adults as are themselves of no value to the community. The laws of nature require the obliteration of the unfit and human life is valuable only when it is of use to the community or race."[59]

As the 1920s progressed, negative eugenics often accompanied racial discourses promising a more auspicious environment for the approaching national palingenesis. As the newly reformed German Society for Racial Hygiene announced in 1922, "the renovation of our whole outlook on life (*Weltanschauung*) is of decisive importance. The welfare of the family, even in distant generations, must be recognised in the vision of all as a higher good than mere personal comfort; and in governmental policies the needs of the present must not obscure the future of our race." In visualising the nation racially, the Society claimed that it pursued the cultivation of racially-aware and healthy citizens. But the proposals advocated by the Society also exemplify how the task of creating a racially conscious culture became intertwined with the desire to purify the national body in accordance with the principles of negative eugenics. To this end, the Society asked for the passing of legislation on the "sterilisation of defective individuals by their own wish or with their consent" and, "in order to prevent the reproduction of anti-social and other very defective individuals, their segregation in labour colonies."[60]

Whilst there is a clear connection between the increased popularity of eugenic technologies of human improvement and a traumatised sense of national identity in most European countries following the First World War, schemes of negative eugenics also appeared regularly – and often more successfully – in countries unaffected by military conflicts, as was the case in Scandinavia.[61] What, then, is the rationale behind this enthusiastic embrace of the negative eugenic repertoire in countries like Sweden, Denmark and Norway? The scientific ethos described in Chapter 1, allowing eugenicists to establish and believe in their role as biological guardians of the nation's body generated a polysemic application of eugenics around Europe. The Swedish psychiatrist Herman Lundborg justified the eugenic involvement in shaping the national community's future in his 1922 *The Threat of Degeneration* thus:

> Of what avail are whole piles of gold, even all the riches in the world, if, for their sake, we head for great disquietude and meet with degeneration in a comparatively short time. It is not easy for an individual to resist all the temptations that are evoked by wealth and luxuries. It is perhaps even more difficult for a whole people to choose a course of self-denial, to do without comfort, to forgo diversions and pleasures and instead live frugally and work hard for the sake of improvement. Where, then, is salvation to be found, one might ask. Surely the whole of mankind is not doomed to destruction? The answer is: We must pay attention to the genotype to a far greater extent than hitherto; that is, we must work far more than is being done now for the lineage and the race, for good families and healthy children.[62]

Using the metaphor of degeneration, Lundborg singled out the compelling need to explain the nation's biological existence through the protection of the individual and the race. Conscripting the eugenic frame to emblematise and legitimise the protection of the nation from degeneration became a regular practice amongst eugenicists. In Romania, the physician Ioan Manliu similarly hoped that sterilisation would provide the antidote to Romania's increased national degeneration. In his 1921 *Fragments of Eugenics and Social Hygiene*, Manliu proposed the following:

> 1) Every degenerate individual should be sterilised and, if possible, returned to society. 2) Every degenerate and sterilised individual should be kept in isolation in asylums and colonies until he/she can be returned to society as a useful member. 3) Only those individuals who still pose a danger to society after their sterilisation should be isolated for life, while they should sustain themselves and society through work in gardens, workshops, etc.[63]

The only way to regain control, he concluded, was the "mass sterilisation of degenerates."[64] In the same vein, the Austrian eugenicist Alois Scholz noted that: "[o]nly if we promote the strong and that which is able to live, and wipe out that which is unable to live, as demanded by nature, are we going to promote racial hygiene."[65] Reading eugenic arguments phrased in such absolutist negative terms substantiates Maria Bucur's assumption about the relationship between eugenicists and the state, namely that "[v]arious forms of individual empowerment came to depend strictly on two fundamental principles: collective (i.e. national) interests, especially of a multigenerational nature, always took precedence over individual ones; and it was the responsibility of the state to protect these interests."[66]

The convergence of these views on negative eugenics in various European countries opened a new round of debates over the components and boundaries of the modernity's capacity for biological change and improvement.[67] The collective desire to overcome the tragedies of war created the need for new national values and moral foundations on which to reconstruct society. Associating science with national efficiency, eugenicists argued for a racial revolution specific to their own countries. This new style of eugenics, one defined by the nation-specific political conditions, was illustrated by

the numerous projects of national renewal and racial protectionism
that emerged during and after the 1930s.

Heavenly Foundations, Rational Planning

If "the secular city of the nineteenth century Darwinians was built
upon substantial if transformed heavenly foundations,"[68] the garden
state of the twentieth century was planned according to the rational
principles of human improvement. Presenting the case for legalising
eugenic sterilisation to the general medical public in 1930, *The Lancet*
raised the issue of the state's welfare, namely that "no one, even by
his own consent, should be allowed to undergo an injury that would
depreciate his value to the State as a fighting man, or as the procrea-
tor of fighting men." Political economy, however, was no match for
political biology as "[e]ven if it be admitted that it is necessary for
the State to safeguard the breeding of warriors, the classes whom
the committee propose to allow to consent to their own sterilisation
are not those from whom such stock can reasonably be expected."
Ultimately, reproduction was not only to be supervised by eugenicists, but its ethical value was subsumed by its biological mission:
"The procreative instinct is the race's most potent weapon against
death, and to fetter its freedom appears at first sight to be a dreadful
thing. It can be argued, however, that there are few things more really
dangerous, either to the race or to the individual, than an unfettered
instinct, and that indiscriminate multiplication in the community is
not less deadly than cancer in the human body."[69]

Until 1933, two issues dominated European debates on eugenic
sterilisation: what social and medical categories would be subjected
to sterilisation, and what legislative forms such policies would
require, namely, whether they would be voluntary or compulsory?
In 1928, the Canton of Vaud in Switzerland authorised the sterilisation of those suffering from hereditary mental diseases and those
deemed feeble-minded.[70] A year later, the Danish Sterilisation Law
was introduced, aimed primarily at "[p]ersons who, on account of
the abnormal strength or nature of their sexual instincts are liable
to commit crimes and whose sexual instincts constitute a danger to
themselves and to the community."[71] Sterilisation was conditioned
by the medical diagnosis provided by a physician and by the consent
of the individual, the family or the legal guardian.

A Committee for Legalising Eugenic Sterilisation was formed in Britain in 1929,[72] and a bill for voluntary sterilisation of mental defectives was introduced to Parliament in 1931.[73] Its main argument rested upon the calculation that the institutionalisation of mental defectives was too expensive compared to the considerably more cost-efficient option of sterilisation. The economic argument loomed large during the 1930s, and not only in Britain. At the First International Congress of Mental Hygiene in 1930, Ernst Rüdin codified eugenics in financially viable terms in order to persuade fellow participants of eugenic sterilisation's significance for the application of modern norms of mental hygiene.[74] In Britain, the campaign was, however, unsuccessful as British eugenicists were divided over which preventive methods was most efficient, sterilisation or institutional segregation.

It was also argued that the general public was not suitably informed of, and therefore persuaded by, the necessity of eugenic sterilisation. The neurologist László Benedek indicated the reluctance with which eugenic sterilisation was discussed in Hungary, remarking that "our land requires a wide-reaching preparation before the conserving humanism will be replaced by the more active health protection of our progeny by the sterilization of cacogenic 'minus-variants.'"[75] Undeterred, Benedek drafted a sterilisation bill in 1932, but the Council of Social Hygiene declined to support it. Nonetheless, Benedek's psychiatrically grounded assertions lent legitimacy to pro-sterilisation propaganda. Declaring that eugenic sterilisation was necessary to ensure the race's qualitative improvement did little to change Christian morality concerning marriage and the individual's private life, two characteristics of the national character Benedek considered to be particularly strong in Hungary.[76]

Implicit in these comments is an aspect that was addressed in Chapter 1, but is worth returning to: Galton's characterisation of eugenics as the "religion of the future." As the increasing ideological emphasis on eugenics was given material form through legislative and policy initiatives, these secular theories of human improvement came into open conflict with the religious dogmas advocated by the main Catholic, Protestant and Orthodox Christian churches in Europe. Eugenic claims of national rejuvenation impacted directly on the carefully orchestrated staging of national identities associated with the church in various European countries. Moreover, eugenics challenged the Church's supremacy over sexual reproduction and marriage.

In most Christian Orthodox societies during the interwar period, it was the church that controlled both the spiritual and the physical body of the nation. When Greek eugenicists, for example, discussed the possibility of introducing eugenic restrictions on marriage and childbearing after the First World War, the Minister of Health and Hygiene Konstantinos Filandros consulted with the Metropolitan Church of Athens. Pointing out that "a series of hygienic measures for protecting our nation from dysgenics and promoting eugenics that are the fundament of every hygienic goal" is needed in Greece,[77] Filandros asked for the Church's approval. A similar attitude towards religious authorities informed the public activities of Bulgarian and Romanian eugenicists. The Bulgarian Law on National Health introduced in 1928 may have stipulated that "[t]he nation and the state suffer considerable material and spiritual damage caused by hereditary defects, as a consequence of marriages concluded between physically or mentally ill persons,"[78] but the law defined health and hygiene in a way that was compatible with Orthodox religious faith.

Eugenicists, in general, did not discourage religious beliefs, and many of them were also practicing Christians, postulating that the state's biological aims should reflect the transcendental aims of the church. The Orthodox theologian and professor at the Theological Academy in Sibiu (Transylvania), Liviu Stan, for instance, developed a racial theology in accordance with both scientific and Christian principles, presenting it as part of the glorious destiny that God had planned for Romanians.[79] Concurrently, the Spanish psychiatrist Antonio Vallejo-Nágera maintained that "[t]he regeneration of the race has to be backed up by the [regeneration of the family institution], because the family constituted in accordance with traditional principles of Christian morality represents a storehouse of social virtues, a bastion against the corruption of the social environment, a sacred depository of traditions." And he added further, "[i]f we reflect a few moments on the bases of the family institution as Catholicism understands them, we will soon be convinced of the solid basis which it provides for the regeneration of the race."[80]

In other circumstances, the church intervened directly in finding a solution to demands for eugenic improvement of society. In Bulgaria, as Gergana Mircheva has argued,

> [w]hen the state authority did not provide for marriage restrictions and prenuptial certificates, eugenicists addressed the Church. Its refusal to

carry out the proposed eugenic undertakings, in turn, was natural in view of canonical doctrine which would not legitimize them. Yet at the same time, the Church did not strictly oppose them. On the contrary, it referred their prospective legalization to the state, as a legitimate source of power and thus, tolerated the strategies and instruments of state biopolitics.[81]

The Romanian eugenicist Ioan Manliu also enlisted the Orthodox Church to contribute to the biological rejuvenation of the Romanian nation. The Church, Manliu suggested, should "use its overwhelming moral authority, declare itself in favour of biological purification and act accordingly."[82] Furthermore, he encouraged the Church to actively engage with the eugenic transformation of Romanian society, arguing that:

> The moment has come for [the Orthodox Church] to take part without delay in this [eugenic] movement, in order to ensure scientifically and biologically the happiness of its believers. If the Church firmly popularizes eugenic ideas and collaborates enthusiastically in their realization, it could provide an invaluable service in our struggle against the degeneration and Asiatization of our race.[83]

Manliu had grasped the essential precondition for any project of social and biological palingenesis to succeed in an Orthodox country: its embrace by the Church. As these countries lacked strong civic traditions and functioning modern bureaucracies, the secular state was supposed to work in harmony (the Byzantine principle of "symphonia") with the Church, symbolising the ultimate synchronisation of religion and government. This politicisation of traditional religions was aptly summarised by the Serbian Patriarch Rosić Varnava when he refuted accusations of "bringing politics into the Church! We do not bring the politics into the Church, but those who have lost reason, patriotism and respect bring poison to the entire national organism. [...] Who else will tell the truth to the people of not the national Holy Church?"[84]

Endeavouring to combine eugenics with religious dogma was, however, stronger in Catholic and Protestant countries. Protestant leaders in Germany, like Johannes Wolff and Hans Harmsen, and the *Innere Mission*, the main protestant welfare organisation, were favourably inclined to eugenics.[85] The social activism propagated by the Protestant Churches was well equipped to embrace eugenic ideas of social and biological improvement. According to its founder Johann Hinrich Wichern, the *Innere Mission's* purpose was to work towards "the voluntary charitable engagement of the *awakened (heilerfülltes) Volk* in order to bring about the Christian and social rebirth of the *heilles Volk*."[86] During the 1930s, the *Innere Mission's* main political activist, the demographer Hans Harmsen, re-enacted these principles to reflect contemporary commitments to the racial revolution prophesised by National Socialists. In 1928, he declared that "[t]he improved registration of the physically and mentally feeble, the numerous army of the mentally ill, cripples, the deaf, the blind and congenital criminals, who are fed and cared for at great expense in asylums, madhouses and prisons, prompted the desire to rid the totality of the nation of these harmful gene pools."[87]

In Hungary, the Reformed Church was particularly dedicated to programmes of racial resettlement meant to strengthen the ethnically Hungarian character of those regions perceived to be either depopulated or under threat by other ethnic groups. As the ethnographer Géza Kiss put it in a letter to his Bishop, László Ravasz: "[t]he ancient and pure Hungarian race, the Reformed community, is on the verge of extinction, and an ugly mix of people is coming for their place from the Gypsies, Romanians, Serbs and Germans."[88] The same racial activism animated sections of the Saxon protestant Church in Transylvania during the 1920s and 1930s. Alfred Csallner, one of the most eugenically inclined Saxon priests, had repeatedly called upon the church to embrace the race-hygienic measures necessary for the racial regeneration of the Saxon nation. The priest was not only charged with his congregations' spiritual well-being, but its biological fitness. The Saxon priest must, according to Csallner, become a "torchbearer" of the eugenic reinvention of the Saxon community. "The Church itself," as Tudor Georgescu convincingly argued, "was to become a *Staatsersatz* of sorts, a substitute for the impossibility of a politically independent and viable Saxon state." Within this environment, "race-hygienic ideologues had become the 'prophets' whose tenets the Church, along with its priests and teachers, were duty bound to spread and manage."[89]

As seen, if the Orthodox Church was supremely ambiguous about its involvement with secular and political movements during the interwar period, the Protestant Church, on the contrary, was actively seeking to accommodate eugenic theories of human improvement within its social and welfare programmes. Neither, officially, opposed negative eugenics. The Catholic Church, on the other hand, was one the institutions that played a significant role in opposing the introduction of negative eugenic policies. What Harry Paul wrote about Catholicism and Darwinism can convincingly be applied to the convoluted relationship between this religious denomination and eugenics: "Although Darwinism per se was never anathematized by the Roman lions of orthodoxy, many of the ideas with which Darwinism was associated were condemned, especially in history and philosophy, the two secular areas Rome regarded as most dangerous."[90] In the case of eugenics, "the domain of concern was reproduction."[91]

It should not be assumed, however, that the attitude of the Catholic Church towards eugenics was simply and unequivocally hostile.[92] In France, for instance, Catholic leaders often looked favourably to eugenics, and the Jesuit René Brouillard noted in 1930 that "[i]n principle, Catholic morality does not condemn all eugenic science."[93] Endorsing this view, Abbot Jean Dermine, professor at the Theological Seminar of Bonne-Espérance (in Belgium) and Monsignor Dubourg, the Bishop of Marseilles, both spoke at the congress on "The Church and Eugenics" organised in 1930 by the Association of Christian Marriage, of family as the embodiment of Christian eugenics, and of "true eugenics," which can only exist in accordance with Christian morality, respectively.[94] Religious dogma conflated with the notion of a eugenically purified family and society.

In Austria, the theologian Johann Ude and the *People's Watch* association he had established in 1917, campaigned for eugenic population policies, describing these as accordingly as "authentic patriotic work" and the expression of "national morality."[95] Although Ude was ambiguous about negative eugenics, he very clearly linked the concept to the new politics emerging after the war, seeing racial improvement as the pillar of the new system for the post-war era. In Germany too, and as Ingrid Richer has demonstrated, Catholics looked affirmatively at positive eugenics deemed necessary after the demographic devastation caused by the First World War. Representatives of the Church often endorsed policies designed to improve society through marriage counselling and hygiene education.[96]

Negative eugenics was also deemed morally and religiously objectionable. As Monika Löscher has noted, "[i]n the 'Catholic milieu', the catalogue of eugenic measures was reduced to a moral appeal to rationality. In this context, abortion, sterilisation and family planning were seen as morally unacceptable. Enlightenment and the education of the coming generation according to eugenic principles (*eugenischen Verantwortung für das kommende Geschlecht*) were limited by the Catholic Church's perception of morality (*Sittlichkeit*)."[97] This Catholic morality and its relationship to eugenics were theologically debated by many religious leaders in Germany and Austria, including Joseph Mayer, Hermann Muckermann and Albert Niedermeyer, and in Hungary, by Tihamér Tóth and József Somogyi.[98]

Amidst the growing acceptance of negative eugenics amongst Christian states in Europe and beyond, Pope Pius XI issued the Encyclical on Christian Marriage, *Casti Connubii* in 1930. The encyclical castigated the prevention of "unworthy" life advocated by eugenicists both as an expression of excessive secularisation and of the state's interference in the individual and family's private sphere.

> But another very grave crime is to be noted, Venerable Brethren, which regards the taking of the life of the offspring hidden in the mother's womb. Some wish it to be allowed and left to the will of the father or the mother; others say it is unlawful unless there are weighty reasons which they call by the name of medical, social, or eugenic "indication." Because this matter falls under the penal laws of the state by which the destruction of the offspring begotten but unborn is forbidden, these people demand that the "indication", which in one form or another they defend, be recognized as such by the public law and in no way penalized.[99]

Yet Catholicism's traditional approaches to contraception, marriage and family life were not seen necessarily in conflict with eugenic teachings: "[w]hat is asserted in favour of the social and eugenic 'indication' may and must be accepted, provided lawful and upright methods are employed within the proper limits." But such acceptance did not condone the termination of life some eugenic programmes argued for: "to wish to put forward reasons based upon them for the killing of the innocent is unthinkable and contrary to the divine precept promulgated in the words of the Apostle: Evil is not to be done that good may come of it."[100]

Condemning contraception and sterilisation as against Christian morality, the encyclical enforced its view of marriage as essential to the healthy functioning of modern society.

> Finally, that pernicious practice must be condemned which closely touches upon the natural rights of man to enter matrimony but affects

also in a real way the welfare of the offspring. For there are some who over solicitous for the cause of eugenics, not only give salutary counsel for more certainly procuring the strength and health of the future child – which, indeed, is not contrary to right reason – but put eugenics before aims of a higher order, and by public authority wish to prevent from marrying all those whom, even though naturally fit for marriage, through hereditary transmission, bring forth defective offspring.[101]

But if some form of eugenics was acceptable to the Catholic Church, it was still targeted for its ambition to replace religious authorities with scientific experts.

Casti connubii was a belated, but powerful, response to Galton's vision of eugenics as "the religion of the future." But it would be a mistake to assume that eugenic dedication was abandoned by the Catholic clergy overall. Indeed, Hermann Muckermann, sensing the importance of contextualising the papal position within his own cluster of political beliefs asserted in his 1934 *Eugenics and Catholicism* that, ultimately, "eugenics can be successfully reconciled with Catholicism. Since Catholicism consistently adheres to a natural ethic, in order to elevate it to the world of the supernatural, it would actually be most surprising to say the least if it were not prepared to accept the well-supported results of eugenic research, and to assimilate them step by step as they appeared."[102] In fact, Muckermann suggested that Catholicism and eugenics were not fundamentally different as in both religion and biology certain absolute principles could be found. This was a renewed affirmation of Muckermann's Catholic beliefs as well as his firm attachment to the growing appreciation of the anthropological project eugenics pursued before culminating in the racial beatitude of the revived national community. "We are all of us," he concluded,

working towards a national eugenics with the same sense of deep responsibility that Galton himself felt for his own people. Different religious sects have different point of view, and each must respect the view of the others. But all should work together for a national eugenics on the basis of a natural ethic. There is no better way for us to serve the future of our people than to work together, consciously and powerfully for the progress of eugenics.[103]

But it was not a new modern Christian morality, but the advent of National Socialism in Germany after 1933 that fully confirmed

Muckermann's belief in a eugenic palingenesis. As we shall see in the next chapter, the quest to create ethnically purged and homogeneous communities embarked upon by so many European countries in the 1940s, whose functions were strictly limited and regulated by the state, not only symbolised the decisive fusion of modernism with eugenics, but also provided the rationale for the ultimate enactment of the biological utopia centred on the nation and race.

4

EUGENICS AND BIOPOLITICS, 1933–1940

The approach to the diffusion of eugenic ideas across the European continent pursued in this book also illuminates the relationship between eugenics and one of its most significant contemporaries: modern biopolitics. The biological reconfiguration of state-wielded powers over the individual was an important consequence of this transformation. Especially after 1933 the boundary between private and public spheres was increasingly blurred, with the idea of collective responsibility for the nation and race dominating both. In Germany, eugenics seemed to have finally found its grandiose role, as the bridge between science and politics. Other European countries and the US were admonished to learn about the practical applications of eugenics by a political regime that claimed to have an answer to the pathological problems represented by Western modernity. It is worth repeating, however, that as the ideology of an innately biological connection between the individual and racial community, sometimes dictating an effacement of the self in relation to the race, something authorising an aggressive attitude towards other members of society categorised as injurious to the race, Nazism was both an assault on eugenics and submission to it. Eugenics emerged not only as a scientific critique of degenerative modernity, or as a process of political, legal and institutional control over the population contained within a delimited territorial space, but also as the expression of a particular race. The identity of any given race was delineated by the boundaries that separated those who belonged to the community from foreigners and outsiders who remained aliens or potential enemies. Prompted by the need to generate a powerful

92

sense of cohesion and shared identity amongst its adherents in the wake of perceivably profound and structural social changes, eugenicists appealed to racial imagery in order to justify their biologisation of national belonging.

The racial eugenic iconography characterising the 1930s and 1940s was quite different from preceding eugenic discourses. In the previous chapter we discussed how and why eugenics had gradually become one of the most potent expressions of the modern scientistic quest for national rejuvenation during the 1920s. Detailed scholarly analyses of Nazi racially and politically charged demographic policies, of which eliminating the Jews was the most barbaric expression, had brought to light how restrictive reproductive legislation and policies have also affected other social groups in Nazi Germany, as well as the lives of women and children.[1] In this chapter, we will not discuss Nazi racial anti-Semitism, Hitler's racial war in Eastern Europe or the Holocaust as these topics are well documented in the current scholarship on the Third Reich.[2] Instead, we will investigate how ideas of biological improvement professed by eugenicists reached their fulfilment in the biopolitical states emerging in Europe between 1933 and 1940.

Practical Applications of Eugenics

Reporting from Germany in 1933, the Romanian philosopher Emil Cioran remarked that "[i]n order to understand the spirit of Germany today, it is absolutely necessary to love everything that is exaggerated, everything that emerges out of an excessive and overwhelming passion, to be enraptured by everything that is [characterized by] irrational élan and disconcerting monumentality." Rather than passively facing its historical destiny, the German nation, Cioran continued, preferred instead to "live a life of mad enthusiasm and admirable effervescence," having the "courage of annihilation, the passion for a fertile and creative barbarism, and especially a messianism that foreigners find difficult to understand."[3] Cioran's infatuation with National Socialism was not unique amongst European modernist intellectuals. Like many others within and outside of Germany, Cioran was impressed by this "reborn Germany," and the Nazis' messianic claim to create "a total culture, not just expressing the genius of the race, but embodying the sacred canopy and underpinning the organic community required to solve the problem of modernity."[4]

On the one hand, the creation of this new Germany presupposed an ideological transformation of the national community, or, to use Peter Fritzsche's salutary expression, to turn Germans into Nazis.[5] On the other hand, however, it promoted a corporeal regeneration of the body politic, the creation of a new man purged of degenerative characteristics and decadent tendencies. As the Nazi ideologue Richard Walther Darré insisted, a new racial nobility would be created through rational selection and breeding: "[e]very available means should be used to achieve the goal that the creative blood in the body of our people, the blood of human beings of the Nordic race, should be preserved and increased, because on this depends the preservation and development of our Germanness."[6] For Gottfried Benn, the expressionist poet, the anthropological transformation brought about by the politics of the "total state" would ultimately lead to the creation of a new German nation "half from mutation and half from breeding."[7]

The regenerative power of eugenics was invoked as one of the main vehicles for this anthropological revolution envisioned by the biopolitical state after 1933 throughout Europe. If in 1929 Eugen Fischer extolled Mussolini as the architect of the rejuvenated Italian nation, by 1931 Fritz Lenz described Adolf Hitler as "the first politician of truly effective influence to make race hygiene a central goal of all politics, and set himself to put that powerfully into effect."[8] It is easy to see why, after decades of public debate and persuasion, eugenicists like Fischer would see in Nazism the long-awaited political opportunity for the practical application of the principles of racial hygiene.

The Law for the Prevention of Progeny with Hereditary Diseases was officially announced on 14 July 1933, and became effective on 1 January 1934. In theory, it stipulated that "anyone with hereditary diseases may be rendered sterile by surgical means, when, according to medical experience, it is highly probable that the offspring of such person will suffer from severe inherited mental or bodily disorders." In practice, however, several medical categories were outlined, namely: 1) hereditary feeble-mindedness; 2) schizophrenia; 3) manic-depressive insanity; 4) hereditary epilepsy; 5) Huntingdon's chorea; 6) blindness; 7) deafness; and 8) severe hereditary malformation. Chronic alcoholics were also subjected to the law. Hereditary

Health Courts were established in 1934 to supervise the implementation of sterilisation.[9] Marie E. Kopp, an observer for the American Committee on Maternal Health, concluded her report by saying that the German eugenic legislation was "a great step ahead as a constructive public health measure, as a method of preventive medicine, and as a contribution to social welfare."[10]

Evaluating the law's importance to the shaping of the National Socialist racial and demographic policy, Walter Gross remarked in 1941 that the "German legislation on the prevention of progeny suffering from hereditary disease is exemplary. It stands out from the efforts made by other countries in this area for the lucidity and pragmatism which makes sterilization dependent on a legally specified medical diagnosis, one that takes no account of biologically speaking totally irrelevant factors such as economic situation, etc."[11] Darré, Gross and other eugenically-motivated Nazi theorists placed consistent emphasis on the firm imbrication of racial politics and national protectionism, and the role of science in the mythologisation of the race in particular.[12] As one contemporary noted in 1937, "[f]or the last five or six years, Germany has undoubtedly stood first amongst the nations as the largest laboratory of eugenic experimentation in existence."[13] With these direct means of state intervention, eugenicists saw many of their ideas of social and biological improvement put into practice. In Nazi Germany, the omnipresence of the state expressed, as Gisela Bock perceptively observed, "the fact that race hygiene was now elevated to the status of an official doctrine of the regime, which proclaimed the need for a 'prevention of worthless life' (*Verhütung unwerten Leben*) and for its eradication (*Ausmerze*) by means of sterilisation."[14] In 1935, two additional racial laws were adopted: the Law for the Protection of the German Blood and German Honour, and the Law for the Protection of the Hereditary Health of the German People. If the former was meant to assure the purity of the race, the latter made marriage counselling mandatory and stipulated the introduction of health certificates before marriage.

It became clear that between 1933 and 1935 Germany's political elite called upon eugenics to implement the population's biological regeneration,[15] and as a contributor to the journal *Eugenical News* put it:

Germany is the first of the great nations of the world to make direct practical use of eugenics. Other nations, at different periods of history, have assumed control of or sought to promote such matters of national greatness as trade, military prowess, agriculture, education or religion. It is a matter for the statesman to try out and for the historian to record whether national perpetuity, the development of a high culture and strong civilization can be promoted more practically by applied eugenics than by the same national emphasis to other lines of culture or phases of civilization.[16]

This was a model that other European countries sought to emulate. In 1932 the Spanish eugenicist Francisco Haro, for instance, proposed a marriage law according to which sufferers from mental disease, epilepsy, congenital feeblemindedness, leprosy, tuberculosis and syphilis were not permitted to marry unless previously sterilised.[17] In Britain, the Departmental Committee on Sterilisation (the so-called "Brock Committee") was appointed in 1932 and submitted a sterilisation bill to the Minister of Health in 1935, before introducing it into Parliament. There were four categories of persons targeted for voluntary sterilisation: a) those with mental defectiveness; b) those who have suffered from mental disorders; c) individuals with "grave physical disabilities deemed to be inheritable" and, finally, d) those "who are deemed to be likely to transmit mental defectiveness or mental disorder or a grave physical disability to a subsequent generation."[18] As with Britain, so too Switzerland placed the heaviest emphasis on the last category of individuals because, as the Swiss psychiatrist Hans Maier explained to the Eugenics Society in London in 1934, "eugenically, this is all the more important because such defectives frequently marry one another and the children produced by them have a still less promising inheritance."[19]

Reports on the German sterilisation law and discussions of its applicability to other national contexts appeared in print in Turkey, Greece, Bulgaria and Romania, including Sadi Irmak's *Heredity and Its Social and Educational Consequences*; Moisis Moisidis's *Eugenic Sterilisation: Principles, Methods, Application*; Ivan Rusev's *Basic Principles of Eugenics (Racial Hygiene)*; and Eugen Petit and Gheorghe Buzoianu's *Sterilisation from Juridical and Surgical Points of View*, all published in 1934. That same year the secretary of the Czech Eugenics Society Bohumil Sekla, discussing the Czech Medical Association in Prague's assessment of the Nazi sterilisation laws, noted that "for our country, also, the need of proper eugenic legislation becomes urgent. The evolution of the population of the country shows that the same contra-selective agencies must be already at work here as they are in the western countries of Europe."[20] The Czech Eugenics Society also considered the issue at its general meeting in 1936, following which it appointed a Committee to draft a sterilisation law.[21] Hungary also contemplated the "adoption of a sterilisation law according to which feebleminded, insane, alcoholic and criminal persons can be sterilized with the consent of the subject or that of the guardian."[22]

Other countries, such as Sweden, emulated the German sterilisation model and introduced compulsory sterilisation of persons suffering from insanity, feeblemindedness and other mental disorders in 1934.[23] Similar legislation followed in Norway (1934), Sweden and Finland (both in 1935), Estonia (1936) and Latvia (1937), allowing for compulsory sterilisations on medical, social and eugenic grounds.[24] The German government was particularly interested in these developments, and even sent a questionnaire to various European eugenics societies to test their commitment to sterilisation:

> Do laws or legal decisions exist with respect to the prevention of heredi-tary diseased offspring, to the encouragement of those hereditarily healthy, and especially of those hereditarily health with many children? [...] What are the reasons for sterilisation? Are they eugenic, medical, social? On what type of decision is sterilisation based: judicial, sanitary policy, voluntary? Is sterilisation performed itinerantly [by mobile sta-tions]? What methods are used? Are those sterilised kept under observa-tion after their release? Do card indexes about sterilisation exist? When was sterilisation introduced, and how many individuals were sterilised by the end of 1934?[25]

The Romanian response was, in this case, "evasive, because, in real-ity, in Romania systematic and co-ordinated measures to encour-age healthy elements and prevent the development of unhealthy ones, anti-socials, etc., had not been introduced."[26] Undoubtedly, Romanian eugenicists, like their German counterparts, believed in the necessity of the biological advancement of the race, but, their rhetorical ambitions did not always correspond to practical accomplishments. Despite intense debates, lecturing and lobbying, Romanian promoters of negative eugenic measures failed to secure the widespread support necessary for a sympathetic government to enact legislation for eugenic sterilisation.

Turkey is another case in point. Prominent Turkish eugenicists, like Sadi Irmak and Fahrettin Kerim Gökay, favoured eugenic sterilisation but doubted its efficiency in Turkey, a country technologically unprepared for such a widespread form of social and biological engineering.[27] Instead, pre-marital examinations were introduced, as institutionalised by the 1930 Public Hygiene Law, prohibiting those with sexually transmitted diseases, leprosy, mental illnesses and tuberculosis from getting married.[28] On the other hand, Estonian eugenicists like the geneticist Theophil Laanes advocated "the promotion of eugenic marriages," not for the elimination of the hereditarily defective, but for "the purpose of having a biologically (if not numerically) stronger army for the defense of the borders in the next Conflagration of Nations."[29] The Hungarian physician Gyula Darányi expressed a similar ambition at the inaugural meeting of the Eugenics Section of the National Foundation for the Protection of the Family in 1939, when he advocated the creation of Institutes of Marital Counselling with the aims to protect and strengthen the moral and biological values of the family.

In other countries – France being a good example – the Nazi model of compulsory sterilisation was viewed with suspicion, as an example of the individual's obliteration by the totalitarian state.[30] The French eugenicist Georges Schreiber raised the issue of state interference in the private spheres, asking if "the State [has] the right to impose an obstacle to bodily integrity for the good of the nation?" Schreiber believed that "[a] nation whose sentiment resents the power of the state to interfere with the individual will not be converted to obligatory sterilisation by reasoning of any kind; this is why sterilisation will probably never become firmly established in France."[31] Schreiber, moreover, and akin to some Italian and Romanian eugenicists, distinguished between a "Latin eugenics" understood to focus more on improving the social environment and education, and an "Anglo-Saxon eugenics" seemingly preoccupied with negative prevention, selective breeding and racial protectionism. At the first International Latin Eugenics Congress convened in Paris in 1937, the Italian eugenicist Corrado Gini expressed this alleged difference between the two branches of international eugenics thus:

> The variety of circumstances in the Latin countries, the balanced nature and the moderation of their ruling classes, a consequence of their more ancient civilisation and, perhaps also, a more pronounced faculty to

detach their judgements from personal interests, allow us to believe that amongst those Latin countries represented, the discussion of eugenic problems will take place in the objective spirit which is the condition of success in scientific works.[32]

How successful the rhetoric of "Latin eugenics" was amongst other participants is forcefully exemplified by the Romanian eugenicist Gheorghe Banu, who elaborated on the dysgenic factors in Romania and the necessity of a practical programme of eugenics. He concluded by expressing his support for the voluntary sterilisation bill adopted by participants of the XIth Congress of Neurology, Psychiatry, Psychology and Endocrinology in 1931, which stipulated the "sterilisation of the hereditary feeble-minded by X-rays or vasectomy. This sterilisation could be performed only on patients who have been interned for at least five years in a mental hospital and only after the advice of a commission of specialists and the consent of the family [have been obtained]."[33]

In 1939 Banu published *L'hygiène de la race*, which offered both a solid theoretical discussion of heredity, and proposed concrete solutions to the race's biological improvement. Regarding negative eugenics, Banu claimed that whilst some of the objections raised by "moralists and the representatives of the Church" were legitimate, the scientific arguments justifying preventive sterilisation were overwhelming. Eugenic sterilisation, by its nature, bore significant implications for the state as it offered a means by which to cut expenditure and re-invest in other public sector areas rather than offer treatment and protection to perceivably dysgenic social groups. Ultimately, preventive sterilisation was, "first and foremost, of biological importance: it concerned the purity and the vital value of the race."[34] The target, in Romania, therefore, was to work towards a programme of biological rejuvenation in which relationships between the individual and the dominant racial community were mutually advantageous.[35]

More importantly, Banu's programme of practical racial hygiene illustrates how volatile the distinction between "Latin" and "Anglo-Saxon" eugenics is. He, like eugenicists in Germany and Scandinavia, embarked on a eugenic quest for comprehensive solutions to social problems centred on the idea of the biopolitical state, seen as the only authority capable of successfully orchestrating the projects of national renewal. The Norwegian eugenicist Jon Alfred Mjøen promoted the same idea at the XIIth Meeting of the International

Federation of Eugenic Organisations held in 1936. Outlining the principles of "a new state" based on "a biological basis," Mjøen demanded that "[i]n the new State the agencies for medicine and State care shall not, as now, help the defective today so that two defectives shall arise tomorrow; it will, on the other hand, simply work to make them superfluous. The new biological State shall, in other words, be built upon the leading principle: We must distinguish between the right to live and the right to give life."[36]

This eugenic transformation of the national body was, as I have explained above, becoming a customary component of the representation of the new biopolitical state that emerged in most European countries by the 1930s. It is in regard to this representation that we must look at another dimension of the relationship between eugenics and biopolitics: racial vocabularies. Eugenicists, like most other intellectuals and politicians engaged in the process of national palingenesis, were amongst the most ardent supporters of racial narratives of identity. "Knowledge of races," claimed Banu in 1940, "demands a racial hygiene. This is eugenics, which forms the basis of the development of any community or nation. Nothing, indeed, can be of greater importance than the investigation and supervision of factors liable to injure the racial, physical and mental qualities of future generations."[37] But to understand what factors may disrupt the nation's racial existence one must first decide what a nation is.

Eugenic Entopia

The myth of national palingenesis was one central concept that eugenics shared with political ideologies like fascism and Nazism; describing the nation in racial terms was another. In the 1921 preamble of the fascist programme, Mussolini described the nation thus: "[t]he nation is not simply the sum of living individuals, nor the instruments of [political] parties for their own ends, but an organism comprised of the infinite series of generations of which the individuals are only transient elements; it is the supreme synthesis of all the material and immaterial values of the race."[38] Fascism, as Mussolini further declared at a fascist congress in Rome in the same year, "must concern itself with the racial problem; [and] fascists must concern themselves with the health of the race."[39]

If Mussolini aligned himself with the idea of national palingenesis during crises experienced by the Italian society during the First World War, its appeal to the leader of the Legionary Movement, Corneliu Zelea Codreanu, was reinforced by the creation of Greater Romania in 1918, and its subsequent nation-building project. As a defender of the Romanian blood and soil, Codreanu spoke of the need to restructure the national mythopoeia according to a new natural ontology. "Harmony," Codreanu believed, "can be re-established only by the reinstatement of natural order. The individual must be subordinated to the superior entity, the national collectivity, which in turn must be subordinated to the nation."[40] The supreme entity, the nation, was the succession of its generations through the centuries, uniting both dead and living members of the national community in a racial continuum: "When we say the Romanian nation, we mean not only all Romanians living in the same territory, sharing the same past and the same future, the same dress, but all Romanians, alive and dead, who have lived on this land from the beginning of history and will live here also in the future."[41]

This form of nationalist mythology of redemption was on one level clearly mystical; but on another it translated a form of scientism characteristic of all fascist movements during this period. "The final aim of the nation," Codreanu persisted, "is not life *but resurrection*."[42] These repeated references to the Christian model of suffering confirm Codreanu's narrative of redemption, in as much as they proclaim the nation's capacity to resurrect itself from its current moral and physical decadence. In forging the new Romanian nation, then, Codreanu did not create something entirely new, but re-enacted and thus reshaped the eternal racial substance he identified as residing in the biological flow uniting Romanians of the past with those of the future.

Attempts to revive national consciousness centred on allegories of race and blood. If the Hungarian racial hygienist Lajos Méhely reiterated the importance of blood purity for "the strict protection of racial borders,"[43] the philosopher Nichifor Crainic affirmed that racial nationalism was both biological and spiritual, and that Romania needed to embark upon her spiritual redemption having already discovered her biological roots. Crainic proposed a corresponding, anthropological, revolution:

The problem of regeneration should be addressed in terms of ethnicity [although] this is discarded and denied by internationalist doctrines.

> [Ethnicity] is not just a general biological concept, but one specifically anthropologic. Man is both body and soul, but he does not come into the world with the body of just another animal and than later adds spirit in order to differentiate him artificially from his animal body. From his birth man is both body and spirit, and together they make the same being. This is both an anthropologic and an ethnic being. The idea of regeneration, as it is conceived of by the new ethnic nationalism, concerns man in its integral, harmonious form, both morally and physically.[44]

Interwar racial anthropology endorsed the view that blood groups were inherited according to Mendelian laws of heredity, thus impregnating the individual with one distinguishing attribute impervious to internal or external influences. As the Italian haematologist Leone Lattes declared in his 1923 *Individuality of the Blood*: "The fact of belonging to a definite blood group is a fixed character of every human being, and can be altered neither by the lapse of time nor by intercurrent diseases."[45]

The 1938 Manifesto of Racial Scientists, written by the Italian racial anthropologist Guido Landra and approved by Mussolini, not only conceptualised the ideological components of Italian fascism in racial terms, but also emphasised the importance attributed to the principle of blood in determining the Italian nation's identity. "There exists today a pure 'Italian race'," stipulated the manifesto, amongst other things; but "[t]his announcement is not based on the confusion of the biological concept of race with the historic and linguistic concept of people and nation, but on the very pure blood relationship that unites Italians of today to the generations that for thousands of years have inhabited Italy. This ancient purity of blood is the grandest title of nobility of the Italian nation."[46]

In his 1940 *Fascism and the Albanian Spirit*, Lazër Radi applied this Italian fascist model of national identity to Albania, arguing it needed to acquire a new racial morality as "one of the many virtues of our people is the particular concern for the upbringing of man, taking specific care that the clearness of the race and maintaining the continuity of the family. I believe no other country believes in the force of blood [as much as Albania]."[47] Radi also insisted on the importance of defending the Albanian race, especially in those remote mountainous regions where vital virtues had been preserved. Ultimately, preserving the race meant ensuring that Albanian racial essence survived, according to which an Albanian was always brave, especially on the battlefield where dying with honour was more important than life itself because "dying with honour brings relief in pain and misery; Honour is more important than death."[48]

Race, blood, soil and martyrs were some of the metaphors these authors regularly used to convey their message of national regeneration. But another component was equally important. In Greece, national regeneration was accompanied by an equally ambitious cultural project: the creation of the "Third Greek Civilisation." As Ioannis Metaxas urged the Greek nation in 1936, "[r]egeneration from a national point of view: because you cannot exist but as Greeks; as Greeks who believe in the power of *Hellenism*, and through it you can develop and create your own civilisation."[49] Metaxas's template for designing this civilisation was a dynamic and teleological vision of Hellenism that would gradually unite the classical Greek culture with the Byzantine Orthodox heritage.

There were also other politicians and thinkers, like the Slovak philosopher Štefan Polakovič and the Norwegian politician Vidkun Quisling who envisioned the regeneration of the nation not within their own borders but within the rising Nazi European Empire. Exalting the creation of the Slovak Republic under the presidency of Monsignor Jozef Tiso in 1939, Polakovič invited "every member of the nation [to] participate in this new order, because it is the dictate of the epoch, which always asks for absolute adjustment in matters of natural rights and natural morality. Those," he warned, "who do not understand the epoch will be overturned and crushed in time."[50] For Quisling, too, the "new Norway must built on Germanic principles, on a Norwegian and a Nordic foundation."[51] Under German occupation, the Czech politician Emanuel Vajtauer similarly welcomed "the rebirth of the idea of a great European collective. We can only welcome the fact that once again people talk of the Reich as a great family of European nations, united by the idea of European cultural tradition and once again setting itself a great common purpose, worth of the former glory of a dynamic continent. Only this development," Vajtauer continued, "can rid us of the anxiety we have never before been able to suppress. Only in this new European unity can we rediscover our lost field of national action."[52]

In a typical biopolitical fashion, these authors portrayed the nation's historical mission as one of constant combat against the injurious legacies of the past, invoking the need for an alternative political order and, ultimately, temporality, both of which based on the rebirth and reinvention of the ethnic community. But this was not the racial future envisioned by Nazi purists and some of their supporters in the Scandinavian countries. Mussolini, Codreanu and Metaxas, as known, did not accept the Nazi concept of racial purity and supremacy even if they believed each nation had a racial nucleus which was permanent. It was this racial nucleus that eugenics and anthropology identified as "Italian," "Romanian" and "Greek," respectively.[53]

The Greek anthropologist and eugenicist Ioannis Koumaris expressed this broad eugenic national philosophy, one based not on racial purism but on syncretism, by formulating the theory of the "fluid constancy" of the race. This meant identifying the various characteristics of the race, their direct influences, and their inter-active effects. Koumaris defined the Greek race as having "almost uniform characteristics, physical and psychical, inherited in its descendants; it has all the principal characteristics of the basic ele-ments, which are all Greek and indigenous in spite of the variety of types."[54] Millennia of ethnic mixing notwithstanding, a Greek racial essence survived miraculously into the modern times. "This race," Koumaris argued,

> is distinguished today by a kind of "fluid constancy", with its own soul and especially with its own variety, dating from prehistoric times. Races exist and will continue to exist; and each one defends itself. Because every infusion of "new blood" is something different and because chil-dren of mixed parents belong to no race, the Greek race, as all others, has to preserve its own "fluid constancy" by avoiding mixture with foreign elements.[55]

If Romanian eugenicists suggested the existence of a "Dacian racial type", which was to be found especially amongst the inhabit-ants of the Carpathian Mountains in Transylvania, Koumaris pos-ited the Greek historical continuity and the permanence of its race on the centre of the classical Greek civilisation, the Acropolis: "[t]he Greek race was formed under the Acropolis Rock, and it is impos-sible for any other to keep the keys of the sacred rock, to which the Greek soul is indissolubly linked."[56] Even the republican and secular discourse of modern Turkish nationalism found the concept of race appealing towards cementing the population's allegiance to a newly reborn Turkey. According to a 1934 biology textbook for secondary education:

> The Turkish race, to which we are proud to belong, has a distinguished place amongst the best, strongest, most intelligent and most competent races in the world. Our duty is to preserve the essential qualities and virtues of the Turkish race and to confirm that we deserve to be members of this race. For that reason, one of our primary national duties is to adhere to the principle of leading physically and spiritually worthwhile lives by protecting ourselves from the perils of ill health, and by applying the knowledge of biology to our lives. The future of our Turkey will depend on the breeding of high valued Turkish progeny in the families that today's youth will form in the future.[57]

To be sure, many prominent political leaders in 1940s Europe, despite their familiarity with heredity theories, never mentioned eugenics as such. But, what Ayça Alemdaroğlu noted on the case of Turkey – namely that "in the eyes of the eugenicists some of [Atatürk's] famous remarks such as 'strong and sturdy generations are the essence of Turkey' and 'the nation should be protected from degenerative perils' were the basis of Turkish eugenics discourse"[58] – certainly applies to other countries as well, not least, of course, Germany, where Hitler's ideas were dogmatically adopted by many eugenicists. In Romania, the general tendency within the government of Marshal Ion Antonescu (1941–44) towards national homogenisation and ethnic purification was consonant with the ideological goals of Romanian eugenics emerging after 1940.

Due to the widespread process of biologisation of national belonging during this period, I have suggested that a continuous interaction between political and eugenic discourses prevailed throughout. Eugenicists anchored their ideas of biological rejuvenation into the general programme of national transformation advocated by political leaders. If the Romanian eugenicist Iordache Făcăoaru assumed that "eugenic ideas, in general, are outlined in the testament of our captain [Corneliu Zelea Codreanu],"[59] the Italian demographer Corrado Gini equally believed that Mussolini's policies were meant to "strengthen the national organism and our country."[60]

These samples of racial characteriology are offered as explorations of local eugenic modernist cultures, through which the boundary between the national and the universal was enacted by virtue of the recurring invocation of a racial collective identity transcending time and space. As one Greek contributor to the *New State* announced: "[t]he Hellenic soul is in harmonious relation with the Hellenic race; [...] the fact that we have been born in a certain place where that race once lived which gave to humanity classical civilization – this is not mere coincidence."[61] Artemis Leontis brilliantly described the relationship between racial imagination, ideas of national rejuvenation and the symbolic geography of the nation as entopia, namely "the aesthetic principle of autochthony. It is the principle of native authenticity. It is the principle that culture is native, that culture is nature, that culture is autochthonous."[62] Eugenic entopia was central to the process of biologisation of national belonging during the 1940s, as were claims to historical and racial continuity. As the Romanian historian Petre P. Panaitescu declared: "We are not only the sons of the earth, but we belong to a great race, a race which is perpetuated in us, the Dacian race. The Legionary movement, which has awakened the deepest echoes of our national being, has also raised 'Dacian' blood to a place of honour."[63]

But, Panaitescu did not nostalgically long for the recreation of the Dacian past. He may have invoked a fantasised ancient empire, like many another in Europe at the time, but looked forward to a racial future defined by autochthonous allegiances to the nation. Similarly, the state-oriented eugenic and racial policies to be enacted throughout Europe in the early 1940s were defined by this autochthonous definition of the dominant ethnic group, seen as the repository of the racial qualities of the nation. When the Romanian eugenicist Făcăoaru declared racial anthropology's unique propensity to determine the *"right to leadership of those superior,"*[64] he not only insinuated that the Romanians were destined to rule over other ethnic minorities, but also that biological, social and political methods were needed in order to assure the dominant group's political and biological leadership. Race was, in this context, employed to legitimate and rationalise the political geometry of a besieged Romanian nation-state.

The minority groups were increasingly relegated to the margins of the dominant group's racial identity, but such a process was not without consequences. Gradually, a counter-narrative emerged to the state's perception of its ethnic minorities. Often, ethnic minorities proposed their own regenerative eugenic programmes, and the adequate means to pursue it. Through their political organisations and churches, as we have seen in the previous chapter, ethnic minorities in various regions of Europe asserted their own eugenic definition of the community, challenging the biopolitical topography of the nation-state created by the dominant group.

Controlling the Ethnic Minorities

Eugenic discourses, as with many of the examples discussed in previous chapters, exclusively reflect the views and demands of the majority of the population, of the dominant ethnic group within the state. Yet, eugenics was articulated by ethnic majorities and minorities alike, as both groups lamented alleged processes of decline, degeneration and extinction. As the biopolitical state presupposed, first and foremost, the creation of an identical ethnic subject, it is worth questioning the extent to which eugenics was employed in diagnosing the ethnic minority's sense of impending crisis and dissolution?

In the Tsarist Empire, for example, fears of national extinction dominated Estonian eugenics since its inception. As Ken Kalling argued, "[f]rom the outset, eugenics was concerned with demography due to high rates of emigration (which became common in the second half of the nineteenth century) and the denationalisation of ethnic Estonians, who adopted the German or Russian language as their own."[65] Saving the race, therefore, depended upon educating Estonians on their own culture as well as the consequences of their denationalisation. If the Estonians experienced educational and political marginalisation under the Russian regime, in Finland – another country which obtained national independence in 1918 – it was the Swedish minority that felt vulnerable and frustrated. The Society for the Promotion of Public Health of the Swedish Speaking Population in Finland was established in 1921. According to one of its main representatives, Harry Federley, the Society had both scientific and practical sections. The goal of "promoting public health," Federley believed, "presupposed both eugenic and euthenic work. The leaders of the society are clearly aware that only through eugenic discipline will they be able to bring about the real improvement of the people."[66]

The Society also issued a proclamation to the Swedish population of Finland, calling for racial protectionism, familial cohesiveness and safeguarding of property. It was a eugenic catechism based on the following commandments: "On our own will, on our innate power ultimately depends the future of the Swedish race in our country. A people mentally and physically strong can hold out even under hard outward pressure!" Or, "[b]e careful therefore when you marry that your chosen one has good health and belongs to a healthy family! Be on your guard against strong drink and sexual diseases! They not only undermine your own health. They can even do harm to your children and their posterity." Not only was the population's health at stake, but its political economy was similarly in danger of being depleted by the combined effects of indifference and external pressure. "Should you be the owner of a piece of ground do not sell it," the Swedish were summoned, "however tempting the offer may be! Rather make it your glory and your pride as your fathers before you to leave it in an improved state to your descendants!" Finally, urban modernity was identified with racial corruption and degeneration: "Above all, do not move into town! There temptations and dangers threaten you, which may be the cause for your own and your children's ruin!"[67] The eugenic frontiers between the Swedish minority and Finnish majority were thus established, indicating how the biological cartography of the nation-state existed as much to invent as to record actual differences between ethnic groups.

The German minorities in Czechoslovakia and Romania provide further compelling examples of minority eugenic discourses during this period. In 1924, the German Association for Local Historical Studies and Local Development was established in Aussig (Ústí nad Labem) whose aim was to strengthen the traditional German identity in the region, but also to articulate new responsibilities for the community's social and racial welfare.[68] By the late 1930s, the leader of the Sudeten Germans, Konrad Henlein, captured the eugenic importance of motherhood and healthy marriages, which to him – like to other nationalists – also constituted an essential component of the pronatalist propaganda among the Sudeten Germans.[69]

Much of the eugenic and racial message was incorporated in the newly founded, rapidly growing societies and institutes for historical and racial research.[70] In 1938, the Saxon priest and eugenicist Alfred Csallner, whom we already encountered, established his National Office for Statistics and Genealogy, whose purpose was not only "to evaluate all statistical matters concerning the German national community in Romania, to alert them to their national duties, and to guard their appropriate application, but, what is more [...] to ensure that all other national- and race-hygienic duties are realized and properly implemented by any other relevant departments."[71] Indeed, in 1940, this National Office was transformed into the Institute for Statistics and Population Policy of the German National Community in Romania, marking the political transformation of the biologically redefined Saxon identity, and its consequent integration into the National Socialist plans for the racial reconstruction of Eastern Europe.[72] Similarly, an Institute for Local Historical Studies was established in 1941 in Käsmark (Kežmarok) accommodating the increased Nazi interest in the racial potential of the Zips and other "Carpathian" Germans living in the independent state of Slovakia.[73]

If the Sudenten Germans and the Saxons in Transylvania embraced radical politics as the natural route to implement their programme of national survival and renewal, others, like the Lusatian Sorbs in Germany and the Csángós in Romania, have absorbed the eugenic narratives that others have projected onto them in order to create their own version of ethno-national identity and national belonging. The Csángó priests Iosif Petru Pal and Ioan Mărtinaş, for example, internalised the dominant Romanian racial narrative created about them, most prominently by the anthropologist Petru Râmneanţu,[74] accepting that the latter's claim that racially the Csángós were, in fact, Romanian. Contrary to other minority groups in Central and Southeastern Europe, especially the Sudeten Germans, the Saxons and the Csángós did not desire the territories' incorporation into the German Reich or Hungary (although many of them relocated outside Romania).

Yet these eugenic and racial discourse espoused by ethnic minorities clearly indicates another aspect of the modernist quest for national palingenesis. In accomplishing the ultimate eugenic goal, a biopolitical state regulated by scientific norms of health and hygiene, eugenicists assumed more than just cleansing the national body of its alleged defective members; they meant, in fact, a complete refashioning of society and the state based on principles of racial homogeneity and protectionism. As the Italian zoologist Marcello Ricci declared in 1938: "[i]t is necessary therefore to recognise that, ultimately, the single greatest benefit for racial improvement could come from the elimination of defectives."[75] In keeping with the reciprocal dependency between the biopolitical regime and its subjects, his compatriot, the physician Giuseppe Lucidi proposed a "census of blood" in 1939, whose objective was twofold:[76]

> 1) *For a scientific-racial end,* an exact study of blood groups, beyond giving documental substance to our racism, would determine the biological characteristics of our race, placing our science at the *avant-garde* of all relevant research, considering that abroad they are actively working while here almost nothing is being done, nor is likely to be done, to develop a solid basis for a racial science. [...]
>
> 2) *For a practical end,* for an optimization beyond the defence of our race, as such research would permit us to know exactly the blood group of any individual, which more than in time of peace, in times of war could save thousands and thousands of lives, or to put it simply, make the practice of blood transfusions more practical.

The Romanian legal expert Ioan Gruia articulated the same racial philosophy in 1940 when he, occasioned by the introduction of anti-Semitic racial laws in Romania, declared: "We consider Romanian blood as a fundamental element in the founding of the nation."[77] In accordance to this position, eugenicists sought to reconcile racial politics with the revolutionary transformation of the state, and to promote a legitimising argument for it which, it was asserted, was the basis for the much needed national rejuvenation. The Hungarian demographer Károly Balás had precisely this transformation in mind when he posed the following question in 1936: "[w]ould it not be worthier of future humanity, advancing in civilization and culture, if the State were to proclaim its right to a regeneration in the interest of the community against the right of destruction?"[78] All this suggests that eugenic biopolitics contained its own contradictory dynamic: it ritualised the importance of the nation but it sacrificed its members for the possibility that the state was born anew.

The Biopolitical State

We have been exploring some of the ways in which eugenic discourses drew on the modernist myths of regeneration, and used them as sources of inspiration for a new form of politics. Whether expressing a majority or minority point of view, eugenicists throughout Europe strove to grasp modernity's world changing dynamic, to appropriate it for their own particular national purposes and transform the chaotic post-1920 environment into new forms of meaning and solidarity and, finally, help their contemporaries become the subjects as well as objects of the biopolitical revolution prophesised by the fascist and Nationalist Socialist regimes.

A month before the introduction of Law for the Prevention of Progeny with Hereditary Diseases, the German minister of Interior Wilhelm Frick reminded racial experts of the importance bestowed on eugenics and biopolitics by the new regime:

> The scientific study of heredity (based on the progress of the last decade) has enabled us clearly to recognise the rules of heredity and selection as well as their meaning for the nation and state. It gives us the right and the moral obligation to eliminate hereditary defectives from procreation. No misinterpreted charity nor religious scruples, based on the dogmas of past centuries, should prevent us from fulfilling this duty; on the contrary it should be considered an offence against Christian and social charity to allow hereditary defectives to continue to produce offspring – having recognised that this would mean endless suffering to themselves and to their kin and future generations.[79]

Casting the German scientists in the role of medical reformers and pillars of the unfolding process of national regeneration, Frick asked for the total commitment to eugenic research: "We must have the courage to rate our population according to its hereditary value, in order to supply the State with leaders. If other nations and foreign elements do not wish to follow us in this course, that is their own affair. I see the greatest aim and duty of the Government of our national revolution in warranting the improvement of our German people in the heart of Europe."[80] Eugenicists and other racial researchers hence voiced not only their pathos for science (as explored in Chapter 1), but equally their therapeutic methods to ensure national improvement (as documented in Chapter 3). It was "no longer a question of theoretic desiderata," the doyen of European racism Georges Vacher de Lapouge wrote to Madison Grant upon reading Frick's lecture, "but of an entire legislation concerning selection, realized through decrees to be applied, and designed to rapidly extinguish undesirable stocks and to perfect the eugenical strains." It was, ultimately, "the birth of a new civilisation, replacing in Germany and soon in the entire world – so we hope – the political ideals and the classic and religious morals, the breakdown of which has upset the social life of all peoples."[81]

To be sure, there was nothing new about this biopolitical use of eugenics. The modernist journal *The New Age* had published an article entitled "Bio-Politics" as early as 1911. Its author, G. W. Harris, defined the term as follows: "a policy which should consider two aspects of the nation: in the first place, the increase of population and competition; in the second place, the individual attributes of the men who are available for filing places of responsibility for the state." The state was summoned to intervene to not only regulate the social selection of worthy individuals, but also to legalise those procedures meant to purify society of its unwanted members. Abortion, for instance, was commended as an efficient way to limit "the production of illegitimate children." Equally radical measures were needed, Harris believed, when dealing with "lunatics and criminal lunatics." Thus, "[u]nless some practical use can be made of the causes of their disease, then a State lethal chamber is the best way out of the difficulty." A new rationality was needed as the foundation of the modern biopolitical state. "Once we do away with the pomp and ceremony and ethical and moral lamentation over death, crime and other evils," Harris concluded, "we shall be able to treat them in a rational way without endeavouring to extract self-satisfaction from the failure of the ungodly."[82]

During the decades following Harris' incipient version of biopolitics, the concept had, however, morphed into a cluster of diverse eugenic theories centring on state's direct intervention into all spheres of life.[83] It is, therefore, appropriate to consider the brand of biopolitics that crystallised by the late 1930s, "not only as a project of elites and experts, but as a complex social and cultural transformation, a discourse – a set of ideas and practices – that shaped not merely the machinations of social engineers, but patterns of social behaviour much more broadly."[84] It is also necessary to take into account that the biopolitical idea of the total state controlling and administering the national organism was an intrinsic component of eugenic discourses across Europe, and not just in fascist Italy and Nazi Germany. As Edward Ross Dickinson aptly noted, biopolitics was a "multifaceted world of discourse and practice elaborated and put into practice at multiple levels throughout modern societies."[85]

Take one of the most convincing discussions of biopolitics during the interwar period, offered by the Romanian eugenicist Iuliu Moldovan in 1926. Entitled *Biopolitics*, this book endeavoured to "militate for a state organisation and activity which are characterised and determined by the supreme duty to assure the biological prosperity of the human capital."[86] The biopolitical state was, therefore, invested with a specific national mission to direct disparate narratives of historical experience and cultural traditions towards the idea of improving the racial qualities of the nation. "The state," Moldovan maintained, "is a political structure, a collective organisation whose aim is to guarantee the utmost biological prosperity of its citizens. This prosperity can only be achieved by respecting integrally the laws of individual and human evolution, in general."[87] The family was placed at the centre of this new biopolitical order grounded in eugenics. Moldovan argued, first, that the family was physically healthy and orderly and, second, that it was socially and spiritually active. More importantly, the relationship between the nation and the state was turned into a specifically crafted form of scientific knowledge, based on heredity and eugenics.

Biopolitics thus operated through investigations of biological processes regulating the triadic relationship between the individual, the nation and the state. In the first half of the twentieth century, biopolitics and eugenics have gathered widespread and enthusiastic support, and numerous eugenicists have worked to systematise the theoretical framework needed for the practical application of biopolitics to the life of nations. Yet, despite sporadic successes, Moldovan concluded, no state has had the power to embark upon such radical a departure from the traditional forms of politics. No later than 1933, Eugen Fischer postulated the same uncertainty about the adoption of biopolitics when contemplating "the new time" inaugurated by the "German ethnic state in its National Socialist form." This was, however, a state founded on the principles of biology, and "the selective and eliminationist hereditary and racial welfare."[88] Had eugenics become national politics? Was the "ethnic state" more than just "an ideal," Fischer asked cautiously? To look at it closely, Fischer concluded, was to confront not a political programme but a form of biological knowledge that was taken to signify "the future and the salvation of our German people and fatherland."[89]

Like eugenics, racial welfare proved essential to the demographic, social and expansionist policies of the biopolitical state. As the Italian minister of education, Giuseppe Bottai explained in 1928: "a state, like the fascist one, which enlarges its social base and extends its roots deeply into the organic mass of the people must necessarily conceive of welfare as a means to preserve the race."[90] Taking this idea further, the president of the Hungarian Association of Biopolitics Lajos Antal corroborated racial welfare with a new political biology he termed "biologism." The only way to achieve the "biological fullness" of the individual and the national community was through biologism, Antal claimed in 1940. Correspondingly, Hungarian biopolitics was based on "the maintenance, care and increase of the biological values of Hungarianness and on the opening of new sources of biological strength." Complementing this biological ideal, Antal also displayed new attitudes towards "the qualitative and quantitative development of the biological values of the Hungarianness" as these were to decide "the future of our nation in the Danube basin." Biopolitics, it was emphasised, was too important to be neglected further; it must be employed to "promote the biological value of Hungarianness," to increase "vitality and biological breeding" of the Hungarian nation. State leadership was summoned to contribute to "the biological future of the Hungarian nation" by introducing economic, social, medical and eugenic policies, thus creating a biologically powerful national body.[91]

That same year, the sociologist Traian Herseni outlined his version of Romanian biopolitics. "With the help of eugenics," he believed, "a nation controls its destiny. It can systematically improve its qualities and can reach the highest stages of accomplishment and human creativity."[92] Herseni subsequently suggested the introduction of biopolitical laws, such as segregation and deportation, as the basis for national regeneration. "The racial purification of the Romanian nation," he alleged, "is a matter of life and death. It cannot be neglected, postponed or half-solved." The scientific language supplied by eugenics was fused with a racist vocabulary: "Without doubt the decay of the Romanian nation is to be attributed to the infiltration in our ethnic group by inferior racial elements; to the contamination of the ancient, Dacian-Roman blood by Phanariot and Gypsy blood, and recently by Jewish blood."[93]

New biological elites, Herseni concluded, rather than social and political institutions, would be the state's foremost biopolitical messengers. In keeping with his scientistic conception of the national life, Herseni reaffirmed his commitment to a eugenic biopolitical programme in unambiguous terms:

Once the change in mentality concerning the physic-racial phenomenon has occurred; once the evaluation and social selection based on racial qualities has been achieved, the most difficult action – but also the most efficient through its qualitative and long-lasting results – must follow: eugenics, which is the improvement of the race through heredity. We need eugenic laws and eugenic practices. Reproduction cannot be left unsupervised. The science of heredity (genetics) demonstrates clearly that human societies have at their disposal infallible means for physical and psychical improvement – but for this to happen there can be no random reproduction and thus the transmission of hereditary defects; and those possessing qualities cannot be left without offspring. Those dysgenic should be banned from reproduction; inferior races should be completely separated from the [Romanian] ethnic group. Sterilisation of certain categories of individuals should not be considered an affront to human dignity: it is a eulogy to beauty, morality and perfection, in general.[94]

The biological definition advocated by eugenics and biopolitical nationalism thus became the norm, rather than the exception in 1940s Europe. This was confirmed by a series of anti-Semitic racial laws introduced in Hungary and Romania between 1938 and 1942, the 1940 Bulgarian Law for the Protection of the Nation, the 1941 Legal Directive on the Protection of the Aryan Blood and the Honour of the Croatian People, the 1942 Law for the Protection of the Race adopted by the regime of Widkun Quilsing in Norway, or the 1942 French Law for the Protection of Mothers and Children. Undeniably, these laws advocated the same political protectionism of the dominant racial group and the exclusion of ethnic minorities, especially Jews, but they also, like in the Bulgarian and French cases, introduced sanitary screening and mandatory premarital examination, aiming at a general biotypological investigation of the nation.

Facing a Second World War, the eugenic rhetoric of the 1940s intensified in its racial tone. Eugenicists in Hungary and in France, for instance, profited from this ideologically auspicious environment to establish racial institutes, like the Hungarian Institute for National Biology (established in 1940) and the French Foundation for the Study of Human Problems (established in 1941). Maintaining the nation's racial potential became of prime political importance alongside instruments for eliminating the "dysgenic groups," be they defined socially or ethnically. The racial mythology, in addition to a whole range of modern hygienic techniques aiming at improving the health of the nation, thus helped create a new political biology whose purpose was to prepare the race, at the expense of ethnic minorities, for the onset of the biopolitical state.

As seen, fusing eugenics with biopolitical projects of biological engineering required the use of historical, linguistic and anthropological arguments as well as medical knowledge. This form of biopolitical knowledge was created equally by historical realities, most prominently by the fascist and Nazi regimes, and by the racial scientists that developed it. Eugenics and biopolitics can thus not be separated from the national cultures they inhabited. And so too the questions of race and nation eugenicists worked with, as well as their interpretations extracted from empirical data and experiments, were shaped by cultural attitudes, social needs and political possibilities. The corollary to this development was that the distinctly modern scientific ethos of eugenics (discussed in Chapter 1) was transformed, during the late 1930s and early 1940s, in a world characterised by a total war, into the biopolitical project of building the perfect ethnic state.

Michel Foucault notoriously defined biopolitics as a modern discipline trying "to rule a multiplicity of men to the extent that their multiplicity can and must be dissolved into individual bodies that can be kept under surveillance, trained, used, and, if need be, punished."[95] Eugenics made it possible to speak of controlling the national body; cultivating and weeding out extraordinary individuals; and purifying the race. The eugenic vocabulary thus overlapped with an adjacent set of fears over racial and national decline. During the 1940s, especially, the eugenic morality invoked during the early decade of the twentieth century succumbed to racial epistemology as positive eugenic ideals were replaced by authoritarian politics. Illustratively, in 1941, Mihai Antonescu, the Romanian deputy prime minister and minister of foreign affairs spoke of the "ethnic and political purification" of the population in Bessarabia and Bukovina, namely "the purification of our nation of those foreign elements foreign to its soul."[96] Yet, there is no documentary evidence to suggest that the Holocaust in Transnistria was dictated by eugenic considerations,[97] thus probing the assumption, most convincingly argued by Zygmunt Bauman, that atrocities committed against the Jews and the Roma during the Second World War can all be reduced to modernist visions of eugenic "gardening."[98]

Yet, what Galton had ultimately predicted did eventually happen. Between 1933 and 1944 many European nations proclaimed "'Jehad,' or Holy War against customs and prejudices that impair[ed] the physical and moral qualities of [the] race."[99] Eugenics had, finally, become a form of biopolitical knowledge which reinstated the supremacy of the laws of nature over culture and with it the subordination of the individual rights and interests to those of the totalising state. Rather than seeing it as leading inexorably to the Holocaust, adopting a perspective grounded in the modernist interpretation of fascism and Nazism as movements of national rejuvenation allows us to see eugenics not as pseudo-science, but as twentieth-century eugenicists saw it, namely a scientistic model of biological and national engineering.

CONCLUSION: TOWARDS AN EPISTEMOLOGY OF EUGENIC KNOWLEDGE

This book has focused on how eugenics emerged and interacted with European national cultures between 1870 and 1940. During this period, eugenics became part of larger social, political and national agendas that included social hygiene, population policies, public health and family planning, as well as racial research on social and ethnic minorities. As I have argued, eugenics widely served as a vehicle for transmitting a social and political message that transcended political differences and opposing ideological camps. Moreover, eugenics was as diverse ideologically as it was spread geographically, and adhered to by professional and political elites across Europe, from West to East, irrespective of their political and cultural contexts.

Three overarching conceptual strategies have guided this discussion of modernism and eugenics: first, the disentanglement of symbolic eugenic geographies, such as the division between Western and Eastern Europe, by looking at national eugenic traditions from a regional and cross-national perspective; second, the introduction of an asymmetric comparison to evaluate different national contexts which shared similar eugenic practices or, in other words, the search for conceptual and ideological meanings in the broader traditions and frameworks of thought in which eugenic texts were produced; and, finally, third, the formulation of a eugenic epistemology, namely that scientific knowledge is a social construct, moulded onto images of society and culture. I have discussed the first two strategies in the Introduction, and I want to briefly elaborate on the third now.

In his *Thinking with History: Explorations in the Passage to Modernism*, one of the most respected historians of Central European modernity, Carl E. Schorske advised new generations of historians to look at neighbouring disciplines for inspiration and encouragement. The history of science, he believed, was one of those disciplines as it "lent itself most readily to a purely internalistic treatment on a progressive premise shared by both the natural scientists and the public." Such academic intrusion is, by no means, unproblematic. As Schorske further noted, "[h]istorians who sought to embed scientific insight in a social matrix were resisted, in part justly, for their inadequate scientific understanding, but also because they tampered with the mythology of the autonomy of science and discovery prevailing in the scientific guild."[1]

Challenging this "mythology of the autonomy of science" is precisely what I have attempted in this volume. Eugenic knowledge was created by the social and intellectual contexts that informed it. Moreover, eugenicists – like other professionals – were frequently enveloped by their social and political existence, and often adhered to dominant social and political practices. Eugenicists were not separated from the culture they inhabited, and the questions they posed about the individual, society and the state, as well as the interpretations they extracted from their empirical data and experiments were shaped by cultural attitudes, social needs and political possibilities. I believe that in order to construct a convincing relationship between modernism and eugenics one must also expose the cultural, social, political and ideological factors that shaped the configuration of eugenic theories.

According to Peter J. Bowler, all scientific theories "have an ideological dimension that must be exposed if we are to understand why these particular ideas about nature were proposed."[2] Recently, such a perspective was forcefully enunciated by Gabrielle M. Spiegel in her reflections on the "Task of the Historian" presented to the annual meeting of the American Historical Association. "If we acknowledge," Spiegel argued,

> that history is the product of contemporary mental representations of the absent past that bear within them strong ideological and/or political imprints – and it seems unlikely that any historian would today disagree with this, whether framed in terms of discourse, social location, or some other form of the historian's fashioning – then it seems logical to include within the determinants of historical practice the impress of individual psychological forces in the coding and decoding of those socially generated norms and discourses.[3]

In attempting to decode various eugenic texts produced between 1870 and 1940 and their ideological ramifications, I came to realise that modernism and eugenics should not be artificially separated

ed together as both have been intrinsic parts of modern
_ıopean history.

This approach entails the task of evaluating the degree and nature
of ideological transfers and of finding a trans-culturally acceptable
set of analytical categories enabling us to address the key components of European eugenic thought as it was formulated on the
basis of their own environment. The relationship between modernism and eugenics, according to this view, is rooted not only in the
Enlightenment myth of human perfectibility, but also in the hereditarian concepts associated with the Darwinian and Mendelian
revolutions in science. But while it reveals much about the interdisciplinary nature of eugenics, this approach should not diminish the
critical role, both theoretical and practical, performed by the modernist modes of conceptualising the individual, society, the nation
and the state, which were not biological.

Through its examination of the eugenic thinking and practices in
Europe between 1870 and 1940, this book dwelt extensively on the
ideological breath and creativity of modernism discussed by Roger
Griffin in his *Modernism and Fascism*. Yet Griffin's broad comparative
and conceptual approach was adopted here with precautions so that
the two overlapping areas of this book's investigation were properly
highlighted: first, the connection between eugenics and theories
of national regeneration and, second, the extensive comparative
framework needed to explain the relationship between modernism
and eugenics. Together with Zygmunt Bauman, Tzvetan Todorov,
Michael Schwartz, Michael Burleigh, Peter Fritzsche, Mark Antliff
and Aristotle Kallis, Griffin has suggested that eugenics should not
be treated as an extraordinary episode in the history of biological
sciences removed from its social, political and national contexts, as
a deviation from the norm which found its culmination in Nazi policies of genocide, but as an integral part of European modernity in
which the state and the individual embarked on an unprecedented
quest for the renewal of an idealised national community.[4]

As I have also argued, the eugenic vision of the biopolitical state
pointed to the creation of an organic society in which social distinctions, divisions between the social and the political, the individual
and the collective would be eliminated, and where biomedical
experts were elevated to the role of defenders of the national community. To substantiate this claim, I have suggested that eugenics
should be seen as a modernist response to the perceived degeneracy

of modernity experienced not just as a cultural, political and social crisis, but also as a biological one. Across Europe, eugenicists excoriated the futility of cultural and political without biological renewal. To accomplish this renewal, it was necessary, however critically, to embrace modern technologies and the modern world, and think in international as well as national terms.

In this sense, the First World War was a defining moment in the crystallisation of European eugenic thinking. By the opening years of the twentieth century, the safety of race and empire had come to loom large in European national obsessions. As seen in Chapter 1, within ruling elites worries about national efficiency were fanned by fears of degeneration and loss of military strength. The First World War broke out in 1914 and, as we discussed in Chapter 2, augmented concerns with the deteriorating health of the civilian population, demographic losses, and the declining birthrates. Eugenics seized the opportunity to become a central element within the newly emerging political cultures forged out of the war. By offering solutions to increasing concerns with the health of the nation, eugenicists hoped to help the state and society through a reconfiguration of its founding myths. Eugenic projects of national regeneration thus located concerns with the race not merely spatially and temporally, but also tried to mobilise individual and collective action, and assist political movements towards the fulfilment of their goals.

By the 1920s, the Austro-Hungarian, German, Russian and Ottoman Empires disappeared altogether, and a plethora of nation-states took their place in Central and Southeastern Europe. For eugenicists, especially in these regions, there were growing opportunities for education, proliferating career openings, and dynamic intellectual and scientific environments. This provided the context for the professionalisation of eugenics to emerge. Being intimately connected to the development of the nation-state did not impede its growth; in fact, the opposite was the case. The need for well-educated, properly trained technocrats furthered eugenicists' self-definition as useful instruments in perfecting society and state. An even more complex process of appropriation of eugenic discourses occurred in those European states characterised by fascist and authoritarian regimes, like Italy, Spain and Portugal. By the early 1930s, as examined in Chapter 3, eugenicists gained influence in public and political spheres across Europe, and played an important role in formulating health laws, organising health departments and institutes of hygiene as well as improving medical systems. Two overlapping conclusions about the interaction between modernism and eugenics during this period thus emerged: first, eugenicists posited science as a solution to the crises facing their countries after the war; and, second, they called for eugenic legislation to prevent the further deterioration of the nation and race.

The same sense of geographical breadth was evident when we examined the internationalisation of eugenics. Eugenicists were collaborative professionals. The notion that eugenicists formed a community transgressing national borders was registered as early as 1912 at the First International Congress of Eugenics held in London, and then reaffirmed at the congresses that followed, culminating in the last, the Fourth International Congress for Racial Hygiene (Eugenics) organised in Vienna in 1940. British, German, French and Italian eugenicists dominated within these international meetings, but the number of Central and Southeastern European eugenicists in attendance grew significantly during the 1920s and 1930s. By meeting regularly, and frequently visiting each other's institutions, the eugenicists mentioned in this book were intimately familiar with the different aspects of eugenic progress in other countries, and sought to provide original solutions to the dilemmas encountered when attempting to apply external eugenic models to their own societies.

The perhaps most convincing illustration of the ideological fusion between modernism and eugenics has come through our discussion of theories of national improvement and the biopolitical state, as attempted in Chapter 4. During the 1940s, national identities in Europe relied heavily on racial typologies and adhered tightly to a selective mythology of glorious pasts. The eugenic and biopolitical writings from this period abound in demands for the containment of the nation's endogenous pathologies as well as its external enemies. As a result, the eugenic and racial stigma placed onto the bodies of social, medical and ethnic communities displaced the locus of national conflict from inside the nation to its vulnerable boundaries. Illuminating the occluded similarities between various racial and eugenic policies should not preclude, however, one from also seeing the important differences amongst European states in terms of implementing and accepting eugenics as a principle of national politics.

In connection to this argument I have tried to argue that there are different histories of eugenics, articulated and shaped by numerous cultural, political and national contexts, by scientific codes and representations of reality whose empirical referentiality differed from country to country. In analysing the eugenic models of modernity these cultures developed from the 1870s until the 1940s, I have highlighted the necessity of proposing a new interpretation for the history of eugenics which takes the radical multiplicity of contexts as well as the complex processes of ideological transmission and reception into account. Substantial comparative research and analytical effort is still necessary to stimulate historiographic interest in these topics from a comparative international perspective. I have here – successfully I hope – only sketched the narrative in such a way so that it reaches out to both academia and the general public, aspiring to create new foundations upon which to build a solid and refreshing approach to these controversial and unsettling episodes in the history of European eugenics.

Between 1870 and 1940, furthermore, a system of eugenic communication developed in Europe whose code we need to understand if our discussion of modernist theories of human improvement is to be enriched conceptually and comparatively. Rather than being mere appendices to specific national traditions, histories of eugenics must be restored to their place within the national histories of European cultures. This recourse to historical memory is essential if, on the one hand, these countries are to be reconciled with their troubled past and if, on the other, the history of interwar eugenic movements is to be systematically analysed through their appropriate local, regional, national and international contexts.

"Eugenics," as Nancy Stepan rightly pointed out, has the "advantage of being contemporary and yet historical: contemporary in that the problems of erecting social policies on the basis of new knowledge in the field of human genetics and reproductive technology are especially pressing today, yet historical in the sense that the eugenics of the pre-1945 period can be viewed as a relatively closed phenomenon of the past on which we can gain some perspective."[5] Indeed, as recent debates on genetic engineering indicate, eugenics has come to serve not only academic purposes. Currently, the term offers a conceptual background for debates on cloning and *in vitro* fertilisation amongst many more. Aware of the general sensitivity surrounding these topics, specialists and lay observers alike have attempted to disassociate themselves from the interwar history of eugenics.[6] Providing that the historiography on eugenics is constantly changing, this book is engaged with the task of evaluating the degree and nature of conceptual transfers of knowledge and ideas, of finding a trans-cultural set of analytical categories as well as new knowledge-production mechanisms. To answer questions about the basic components of European eugenic thought, one must then formulate them on the basis of a regional and trans-regional comparative analysis. But this approach also necessitated the posing of two other questions: first, how far can one go in inserting specific historical experiences and analytical categories into European circulation? And second, how can one test the value of the interpretative models linked to such notions as race and nation in different historical contexts? By examining a diverse range of eugenic ideas and authors as well as by linking state-oriented eugenics with church or minority perspectives, I hope to have convincingly answered these questions, as well as revealed the extent to which eugenic philosophies of identity based on the modern biological knowledge of heredity and evolution have influenced the nature and practice of national politics.

Eugenics was furthermore understood as a cluster of social, biological and cultural ideas, centred on the redefinition of the individual and the national community according to the laws of natural selection and heredity. As such, eugenics promoted a regenerative programme of the nation seeking to counter the negative consequences resulting from the alleged social and biological degeneration of the body politic. In the majority of cases, eugenics relied on the state's intervention to assure the success of its programme of biological rejuvenation. Two directions were generally followed: a) discouraging those individuals categorised as "inferior" from reproducing and b) encouraging those deemed "superior" to value their hereditary importance for the general health of the nation. This state-oriented eugenics, therefore, defined a dominant ethnic group as the repository of national racial qualities while pursuing biological, social and political means to assess and eliminate the factors contributing to its degeneration.

To an extent that has not yet been fully acknowledged, there was also a counter-narrative to the state's perception of its ethnic minorities. Churches, especially the Protestant ones, and ethnic minorities benefited from the popularisation of eugenics in order to propose their own modernist regenerative agenda, and the means to pursue it. Some ethnic groups, like the Transylvanian Saxons in

Romania, embraced radical politics as the natural route to implementing their programmes of national survival and renewal; others, like the Csángós in Romania and Hungary or the Sudeten Germans in Czechoslovakia, have absorbed the eugenic narratives that others have projected onto them in order to create their own version of ethno-national identity and national belonging. Therefore, whilst most of the existing scholarship on eugenics in interwar Europe has focused on nation-states, this book has also revealed its impact on ethnic minorities, offering a new point of departure and comparative basis for how and why ethnic minorities in Central and Southeastern Europe became a part of eugenic reconfiguration of identity at both national and international levels. I can only hope that this book will provoke other scholars to attempt to reconstruct how a wider regional and international diffusion of eugenic ideas and their implementation was possible in Europe and beyond; and in some instances, how these ideas were obstructed and ultimately rejected.

As we celebrated the bicentennial of Charles Darwin's birth in 2009 it may be appropriate to finish this book with the words of a scientist who was revered by eugenicists and their detractors alike, and whose impact on the modern scientific imagination is still unrivalled. "It seems to me," Darwin surmised in his 1871 *The Descent of Man*,

> that man with all his noble qualities, with sympathy which feels for the most debased, with benevolence which extends not only to other men but to the humblest living creature, with his god-like intellect which has penetrated into the movements and constitution of the solar system – with all these exalted powers – Man still bears in his bodily frame the indelible stamp of his lowly origin.[7]

These reflections superbly illustrate the dense creative frameworks into which eugenic theories of human improvement were later to be absorbed. Any attempt, therefore, to recapture how twentieth-century eugenicists imagined and experienced their ideas of biological and national improvement must inevitably begin (and conclude) by understanding how ideas of modern science like evolution and heredity have battled traditional forces, like religion, for supremacy over the human body. At the heart of all these conflicts there was, in fact, a characteristically scientistic attitude, one whose various incarnations I have explored in this book, and which Francis Galton described as "the religious significance of the doctrine of evolution."[8] When Galton spoke of eugenics as the "new religion of the future," he not only hoped to convert the next generations to the new scientistic faith, but also that these new converts would establish eugenics as a universally recognised science. In no other period during the twentieth century were his hopes nearer to fruition than in the wake of the most devastating event in modern history, the Second World War.

NOTES

INTRODUCTION: CONTEXT AND METHODOLOGY

1 Some historiographic reviews of the ever increasing eugenic scholarship include Robert A. Nye, "The Rise and Fall of the Eugenics Empire: Recent Perspectives on the Impact of Biomedical Thought in Modern Society," *The Historical Journal* 36, 3 (1993): 687–700; Frank Dikötter, "Race Culture: Recent Perspectives on the History of Eugenics," *The American Historical Review* 103, 2 (1998): 467–78; Peter Weingart, "Science and Political Culture: Eugenics in Comparative Perspective," *Scandinavian Journal of History* 24, 2 (1999): 163–77; Paul Crook, "American Eugenics and the Nazis: Recent Historiography," *The European Legacy* 7, 3 (2002): 363–81; and Marius Turda, "Recent Scholarship on Race and Eugenics," *The Historical Journal* 51, 4 (2008): 1115–24.

2 The most impressive collective effort to date is Alison Bashford, Phillipa Levine, eds., *The Oxford Handbook of the History of Eugenics* (New York: Oxford University Press, 2010). Previous works include Mark B. Adams, ed., *The Wellborn Science. Eugenics in Germany, France, Brazil, and Russia* (Oxford: Oxford University Press, 1990); Stefan Kühl, *The Nazi Connection: Eugenics, American Racism and German National Socialism* (New York: Oxford University Press, 1994); Nils Roll-Hansen, Gunnar Broberg, eds., *Eugenics and the Welfare State: Sterilization Policy in Norway, Sweden, Denmark, and Finland*, 2nd ed. (East Lansing: Michigan State University Press, 2005) [first edition 1997]; and Marius Turda, Paul J. Weindling, eds., *Blood and Homeland: Eugenics and Racial Nationalism in Central and Southeast Europe, 1900–1944* (Budapest: Central European University Press, 2007).

3 Most convincingly by authors like Laura Doyle, *Bordering on the Body: The Racial Matrix of Modern Fiction and Culture* (New York: Oxford University Press, 1994); Donald J. Childs, *Modernism and Eugenics: Woolf, Eliot, Yeats, and the Culture of Degeneration* (Cambridge: Cambridge University Press, 2001); and Angelique Richardson, *Love and Eugenics in the Late Nineteenth Century: Rational Reproduction and the New Woman* (Oxford: Oxford University Press, 2003).

4 Roger Griffin, *Modernism and Fascism: The Sense of a Beginning under Mussolini and Hitler* (Basingstoke: Palgrave Macmillan, 2007), 62.

5 For a similar approach see Peter Weingart, "Eugenics – Medical or Social Science?" *Science in Context* 8, 1 (1995): 197–207; and Lene Koch, "Past Futures: On the Conceptual History of Eugenics – a Social Technology of the Past," *Technology Analysis & Strategic Management* 18, 3/4 (2006): 329–44.

6 Lene Koch, "Eugenic Sterilisation in Scandinavia," *The European Legacy* 11, 3 (2006): 308.

7 See Melvin Richter, *The History of Political and Social Concepts: A Critical Introduction* (Oxford: Oxford University Press, 1995), 10. See also Reinhardt Koselleck, *Futures Past: On the Semantics of Historical Time* (Cambridge, Mass.: MIT Press, 1985).

8 Nye, "The Rise and Fall of the Eugenics Empire," 688.

9 For Poland, see Magdalena Gawin, *Rasa i nowoczesność: historia polskiego ruchu eugenicznego, 1880–1952* (Warsaw: Wydawnicwo Neriton, 2003); for Hungary, see Marius Turda, *A Healthy Nation: Eugenics, Race and Biopolitics in Hungary, 1904–1944* (Budapest: Central European University, forthcoming 2011); for Romania, see Maria Bucur, *Eugenics and Modernization in Interwar Romania* (Pittsburgh: University of Pittsburgh Press, 2002).

10 Alison Bashford, *Imperial Hygiene: A Critical History of Colonialism, Nationalism and Public Health* (Basingstoke: Palgrave Macmillan, 2004), 5.

11 For a persuasive discussion of this aspect, see Aristotle Kallis, *Genocide and Fascism: The Eliminationist Drive in Fascist Europe* (New York: Routledge, 2009), esp. 48–84 and Marius Turda, *The Idea of National Superiority in Central Europe, 1880–1918* (New York: Edwin Mellen, 2005).

12 Michel Foucault, *Discipline and Punish: The Birth of the Prison* (New York: Vintage, 1977), 26.

13 Reinhardt Koselleck, *The Practice of Conceptual History: Timing History, Spacing Concepts* (Stanford: Stanford University Press, 2002), 152. See also Peter Fritzsche, *Stranded in the Present: Modern Time and the Melancholy of History* (Cambridge, Mass.: Harvard University Press, 2004).

14 Griffin, *Modernism and Fascism*, 181.

15 Havelock Ellis, *The Problem of Race-Regeneration* (London: Cassell, 1911), 71.

16 David G. Horn, *Social Bodies: Science, Reproduction, and Italian Modernity* (Princeton: Princeton University Press, 1994), 4.

17 Emilio Gentile has provided some of the most convincing arguments in favour of this interpretation. See, for example, his *The Struggle for Modernity: Nationalism, Futurism, and Fascism* (Westport, CT.: Praeger, 2003).

18 See Lene Koch, "The Meaning of Eugenics: Reflections on the Government of Genetic Knowledge in the Past and Present," *Science in Context* 17, 3 (2004): 315–31; Merryn Ekberg, "The Old Eugenics and the New Genetics Compared," *Social History of Medicine* 20, 3 (2007): 581–93; Paul Crook, "The New Eugenics? The Ethics of Bio-Technology,"

Australian Journal of Politics and History 54, 1 (2008): 135–43; Nancy E. Hansen, Heidi L. Janz, Dick J. Sobsey, "21st Century Eugenics?" *The Lancet* 372, supplement 1 (2008): 104–7; and Aviad E. Raz, "Eugenic Utopias/ Dystopias, Reprogenetics, and Community Genetics," *Sociology of Health and Illness* 31, 4 (2009): 602–16.

1 THE PATHOS OF SCIENCE, 1870–1914

1 José Ortega y Gasset, "History as a System," in Raymond Klibansky and H. J. Paton, eds., *Philosophy and History* (London: Oxford University Press, 1936), 313.

2 Alexis Carrel, *Man, the Unknown* (London: Hamish Hamilton, 1935), 257.

3 F. A. von Hayek, "Scientism and the Study of Society," *Economica* 9, 35 (1942): 269.

4 Eric Voegelin, "The Origins of Scientism," *Social Research* 15, 4 (1948): 489.

5 See Tzvetan Todorov, *Imperfect Garden: The Legacy of Humanism* (Princeton: Princeton University Press, 2002), 23.

6 Richard G. Olson, *Science and Scientism in Nineteenth-Century Europe* (Champaign, IL.: University of Illinois Press, 2008), 1.

7 Havelock Ellis, *The Problem of Race-Regeneration* (London: Cassell, 1911), 51.

8 E. Ray Lankester, *Degeneration: A Chapter in Darwinism* (London: Macmillan, 1880), 62.

9 It is Durkeim's theory of religion that promoters of the concept of political religion had found useful for their explorations of modern secular ideologies. See Stanley G. Payne, "On the Heuristic Value of the Concept of Political Religion and its Application," *Totalitarian Movements and Political Religions* 6, 2 (2005): 163–74; and Roger Griffin, Robert Mallett, John Tortorice, eds., *The Sacred in Twentieth-Century Politics* (Basingstoke: Palgrave Macmillan, 2008).

10 Peter J. Bowler, *Reconciling Science and Religion: The Debate in Early-Twentieth Century Britain* (Chicago: University of Chicago Press, 2001), 7.

11 Michael Burleigh, "Eugenic Utopias and the Genetic Present," *Totalitarian Movements and Political Religions* 1, 1 (2000): 64.

12 Aaron Gillette, *Eugenics and the Nature-Nurture Debated in the Twentieth Century* (New York: Palgrave Macmillan, 2007).

13 Francis Galton, "Hereditary Character and Talent," *MacMillan's Magazine* 12, 70 (1865): 319.

14 Ibid., 322.

15 Francis Galton, "Heredity Improvement," *Fraser's Magazine* 7, 37 (1873): 116.

16 Ibid., 123.

17 Ibid., 129–30.

18   Angelique Richardson, *Love and Eugenics in the Late Nineteenth Century. Rational Reproduction and the New Woman* (Oxford: Oxford University Press, 2003), 81.

19   L. Hirschfeld and H. Hirschfeld, "Serological Differences between the Blood of Different Races," *The Lancet* 197, 2 (1919): 675–9. See also William H. Schneider, "Chance and Social Setting in the Application of the Discovery of Blood Groups," *Bulletin of the History of Medicine* 57 (1983): 545–62; and Pauline M. H. Mazumdar, "Blood and Soil: The Serology of the Aryan Racial State," *Bulletin of the History of Medicine* 64 (1990): 187–19.

20   P. C. Mitchell, "Preface" to Elie Metchnikoff, *The Nature of Man. Studies in Optimistic Philosophy* (London: G. P. Putman's Sons, 1903), ix.

21   The description belongs to Daniel J. Kevles. See his highly influential *In the Name of Eugenics: Genetics and the Uses of Human Heredity* (Cambridge, Mass.: Harvard University Press, 1985).

22   Francis Galton, *Memories of My Life*, 2nd edn. (London: Methuen, 1908), 290.

23   Francis Galton, *Inquiry into Human Faculty and Its Development* (London: Macmillan, 1883), 17.

24   Ibid.

25   Alfred Ploetz, *Grundlinien einer Rassen-Hygiene. Die Tüchtigkeit unserer Rasse und der Schutz der Schwachen* (Berlin: S. Fischer, 1895), 13.

26   Francis Galton, "Eugenics: Its Definition, Scope and Aims," *The American Journal of Sociology* 10, 1 (1904): 5.

27   Ibid.

28   Francis Galton, "Studies in Eugenics," *The American Journal of Sociology* 11, 1 (1905): 11.

29   Ibid., 25.

30   Peter Gay, *Modernism: The Lure of Modernism from Baudelaire to Becket and Beyond* (London: William Heinemann, 2007), 28.

31   John M. Coulter, the chair of the Botany Department of the University of Chicago, hoped to convince Christian organisations to "add the practical suggestions of biology to their own great motive, and to transform eugenics so that it may really be another effective form of religion." John M. Coulter, "What Biology Has Contributed to Religion," *The Biblical World* 41, 4 (1913): 223.

32   Maximilian A. Mügge, *Eugenics and the Superman. A Racial Science and a Racial Religion* (London: Eugenics Education Society, 1909), 10. Dan Stone has convincingly described the influence Nietzsche had on the British eugenicists. See his *Breeding Superman: Nietzsche, Race and Eugenics in Edwardian and Interwar Britain* (Liverpool: Liverpool University Press, 2002).

33   Caleb Williams Saleeby, *Parenthood and Race Culture: An Outline of Eugenics* (London: Cassell, 1909), ix.

34 Ibid., 304.

35 Edgar Schuster, *Eugenics* (London: Collins, 1912), 255.

36 *Internationalen Gesellschaft für Rassen-Hygiene* (Naumburg: Lippert, 1910), 4–6.

37 Francis Galton, "A valószínűuség, mint az eugenetika alapja," *Huszadik Század* 8, 12 (1907): 1013–29. An earlier example is Galton's 1904 text on the definition of eugenics which was translated in the *Archiv für Rassen- und Gesellschaftsbiologie* in 1905 and then commented upon in the Hungarian journal *Huszadik Század* in 1906.

38 Francis Galton, "Probability, The Foundation of Eugenics," in Francis Galton, *Essays in Eugenics* (London: Eugenics Education Society, 1909), 98–9.

39 Max Nordau, *Degeneration* (Lincoln: University of Nebraska Press, 1993), v. [first edition 1892]

40 Ibid., 2.

41 Roger Griffin, "Modernity, Modernism, and Fascism. A 'Mazeway Resynthesis'," *Modernism/Modernity* 15, 1 (2008): 21, n. 5.

42 Lankester, *Degeneration*, 32 [emphasis in the original]

43 Giuseppe Sergi, *Le Degenerazioni Umane* (Milan: Fratelli Dumolard, 1889), 25.

44 Sheila Faith Weiss, *Racial Hygiene and National Efficiency: The Eugenics of Wilhelm Schallmayer* (Berkeley: University of California Press, 1987), 3.

45 Wilhelm Schallmayer, *Über die drohende körperliche Entartung der Kulturmenschheit und die Verstaatlichung des ärztlichen Standes* (Berlin: Heuser, 1891). The second edition was republished as *Über die drohende physische Entartung der Culturvölker* (Berlin: Heuser, 1895).

46 See Alfred Ploetz, Ernst Rüdin, "Der Alkohol im Lebensprozeß der Rasse," in *Bericht über den IX. Internationalen Kongress gegen den Alkoholismus (Bremen 14–19.IV.1903)* (Jena: Verlag von Gustav Fischer, 1904), 70–107.

47 Karl Pearson, *The Problem of Practical Eugenics* (London: Dulau, 1909), 34 [emphasis in the original]

48 Ibid.

49 Sergi, *Le Degenerazioni*, 228.

50 Dr Madrazo, *Cultivo de la especie humana. Herencia y educación – Ideal de la vida* (Santander: Blanchard y Arce, 1904), 5 and 137. Both quotes are from Richard Cleminson, *Anarchism, Science and Sex. Eugenics in Eastern Spain, 1900–1937* (Bern: Peter Lang, 2000), 81–2.

51 Pearson, *The Problem of Practical Eugenics*, 37.

52 See the discussion in Lesley A. Hall, "Malthusian Mutations: The Changing Politics and Moral Meanings of Birth Control in Britain," in Brian Dolan, ed., *Malthus, Medicine, and Morality: 'Malthusianism' after 1798* (Amsterdam: Rodopi, 2000), 1–63.

53 Sidney Webb, *The Decline in the Birth-Rate* (London: Fabian Society, 1907) and Ethel Elderton, *Report on the English Birth Rate. Part 1. England North of the Humber* (London: Dulau, 1914).

54 See Richard A. Soloway, *Demography and Degeneration. Eugenics and the Declining Birthrate in Twentieth-Century Britain* (Chapel Hill: The University of North Carolina Press, 1990).

55 Daniel Pick, *Faces of Degeneration: A European Disorder, c. 1848–1918* (Cambridge: Cambridge University Press, 1989), 2. See also J. Edward Chamberlain, Sander L. Gilman, *Degeneration: The Dark Side of Progress* (New York: Columbia University Press, 1985); and Stephen Arata, *Fictions of Loss in the Victorian Fin-de-Siècle* (Cambridge: Cambridge University Press, 1996).

56 David G. Horn, *The Criminal Body: Lombroso and the Anatomy of Deviance* (London: Routledge, 2003).

57 See, for example, Emil Mattauschek, "Einiges über die Degeneration des bosnisch-herzegowinischen Volkes," *Jahrbücher für Psychiatrie und Neurologie* 29 (1908): 134–48.

58 Robert Reid Rentoul, *Race Culture; or, Race Suicide? A Plea for the Unborn* (London: Walter Scott Publishing, 1906), xii.

59 Ellis, *The Problem of Race-Degeneration*, 69.

60 See Weindling, *Health, Race and German Politics*, 63–4.

61 Ibid., 89.

62 G. Sergi, *La Decadenza della Nazione Latine* (Milano: Fratelli Bocca, 1900).

63 See Mark Antliff, *Avant-Garde Fascism: The Mobilization of Myth, Art, and Culture in France, 1909–1939* (Durham: Duke University Press, 2007).

64 Quoted in Angus McLaren, "Reproduction and Revolution: Paul Robin and Neo-Malthusianism in France," Dolan, ed., *Malthus, Medicine, and Morality*, 170. See also Elinor A. Accampo, "The Gendered Nature of Contraception in France: Neo-Malthusianism, 1900–1920," *Journal of Interdisciplinary History* 34, 2 (2003): 235–62; and Richard Sonn, "'Your Body is Yours': Anarchism, Birth Control, and Eugenics in Interwar France," *Journal of the History of Sexuality* 14, 4 (2005): 415–32.

65 Luis Bulffi, "Dos palabras," *Salud y Fuerza* 1 (1904): 1–2. Quoted in Cleminson, *Anarchism, Science and Sex*, 134,

66 S. [Alexandru Sutzu], "Evoluțiunea și hereditatea," *Gazeta medico-chirurgicală a spitalelor* 5, 12 (1874): 182–7.

67 Mihail Petrini-Galatzi, *Filosofia medicală: Despre ameliorațiunea rasei umane* (Bucharest: Tipografia D. A. Laurian, 1876).

68 Mladen Jojkić, *Pokušaj fiziološko-patološke studije o srpskom narodu* (Subotica: Štamparija Vinka Blesića, 1895).

69 Břetislav Foustka, *Slabí v lidské společnosti. Ideály humanitní a degenerace národu* (Prague: Jana Laichtera, 1904).

70 Bretislav Foustka, *Die Abstinenz als Kulturproblem mit besonderer Berücksichtigung der österreichischen Völkerstämme* (Vienna: Verlag von Brüder Suschitzky, 1908), 6.

71 C. W. Saleeby, *The Methods of Race-Regeneration* (London: Cassell, 1911), 45.

72 The persistence of the Lamarckian tradition in French biology, for instance, prompted scholars to argue for different version of eugenics in France, characterised mostly by the improvement of environmental influences on mothers and infants. See William H. Schneider, *Quality and Quantity: The Quest for Biological Regeneration in Twentieth-Century France* (Cambridge: Cambridge University Press, 1990).

73 As illustrated by Auguste Forel, *Malthusianism oder Eugenik?* (Munich: Verlag von Ernst Reinhardt, 1911).

74 Heinrich Siegmund, *Zur sächsischen Rassenhygiene* (Hermannstadt: Peter Drotleff, 1901).

75 *The Life, Letters and Labours of Francis Galton*, vol. IIIa, ed. by Karl Pearson (Cambridge: Cambridge University Press, 1930), 222.

76 Alfred Ploetz, "Die Begriffe Rasse und Gesellschaft und die davon abgeleiteten Disciplinen," *Archiv für Rassen- und Gesellschaftsbiologie* 1, 1 (1904): 1–27. A decade later, Fritz Lenz was still troubled by claims that racial hygiene and eugenics were merely subdivisions of social hygiene. He argued that racial hygiene includes both individual and social hygiene, and brings together the qualitative and quantitative demography. See Fritz Lenz, "Zum Begriff der Rassenhygiene und seiner Benennung," *Archiv für Rassen- und Gesellschaftsbiologie* 11, 4 (1915): 445–8.

77 Alfred Grotjahn, *Soziale Pathologie: Versuch einer Lehre von den sozialen Beziehungen der menschlichen Krankheiten als Grudlage der soziale Medizin und der sozialen Hygiene* (Berlin: Hirschwald, 1912).

78 Reproduced in C. B. S. Hodson, "Eugenics in Norway," *The Eugenics Review* 27, 1 (1935): 42.

79 "Eugenic Research in Bohemia," *The Journal of Heredity* 7, 2 (1916): 157; and "Eugenics in Austria," *The Eugenics Review* 5, 4 (1914): 387.

80 Géza von Hoffmann, "Ausschüsse für Rassenhygiene in Ungarn," *Archiv für Rassen- und Gesellschafsbiologie* 10, 6 (1914): 830–1.

81 "An Italian Eugenics Committee," *The Eugenics Review* 5, 4 (1914): 387.

82 Max von Gruber, Ernst Rüdin, *Fortpflanzung, Vererbung, Rassenhygiene* (Munich: J. F. Lehmanns, 1911).

83 Ibid., 122. For a discussion between Ploetz's concept of racial hygiene and Galton's eugenics see Marius Turda, "Race, Science and Eugenics in the Twentieth Century," in Alison Bashford, Phillipa Levine, eds., *The Oxford Handbook of the History of Eugenics* (New York: Oxford University Press, 2010), 98–127.

84 See Marius Turda, "Eugenics, Race and Nation in Central and Southeast Europe, 1900–1940: A Historiographic Overview," in Marius Turda, Paul J. Weindling, eds., *'Blood and Homeland': Eugenics and Racial Nationalism in Central and Southeast Europe, 1900–1940* (Budapest: Central European

University Press, 2007), 1–22; and idem, "'A New Religion': Eugenics and Racial Scientism in Pre-World War I Hungary," *Totalitarian Movements and Political Religions* 7, 3 (2006), 303–25.

85 József Madzsar, "Gyakorlati eugenika," *Huszadik Század* 21, 2 (1910): 115–17.

86 Ibid., 116.

87 Ibid.

88 "A fajnemesítés (eugenika) problémái," *Huszadik Század* 23, 12 (1911), 700.

89 Ibid., 701.

90 Ladislav Haškovec, "Moderne eugenische Bewegung," *Wiener Klinische Rundschau* 26, 39 (1912): 609–11; 26, 40 (1912): 625–7 (part III); 26, 40 (1912): 643–5 (part IV) and 26, 42 (1912): 659–61 (final part).

91 Karl Pearson, *The Academic Aspect of the Science of National Eugenics* (London: Dulan, 1911), 4.

92 Ibid., 27.

93 Major Leonard Darwin, "Presidential Address," *Problems in Eugenics, vol. 1. Papers Communicated to the First International Eugenics Congress held at the University of London, July 24th to 30th, 1912*, vol. 1 (London: The Eugenics Education Society, 1912), 6.

94 Frederic Houssay, "Eugenics, Selection and the Origin of Defects," in ibid, 158–61 and Adolphe Pinard, "General Consideration upon 'Education before Procreation'," in ibid. 458–9.

95 Agnes Bluhm, "Eugenics and Obstretics," in ibid. 387–95 and Alfred Ploetz, "Neo-Malthusianism and Race Hygiene," in *Problems in Eugenics. Papers Communicated to the First International Eugenics Congress held at the University of London, July 24th to 30th, 1912*, vol. 2 (London: The Eugenics Education Society, 1913), 183–9.

96 Ploetz, "Neo-Malthusianism and Race Hygiene," 189.

97 Bleeker van Wagenen, "Preliminary Report of the Committee of the Eugenic Section of the American Breeders' Association to Study and to Report on the Best Practical Means for Cutting off the Defective Germ-Plasm in the Human Population," in *Problems in Eugenics*, vol. 1, 460–79.

98 See, for example, Kurt Goldstein, *Über Rassenhygiene* (Berlin: Julius Springer, 1913).

99 For an analysis of some of the most important complexities of this relationship see, in particular, Diane B. Paul, *Controlling Human Heredity: 1865 to the Present* (Atlantic Highlands, NJ: Humanities Press, 1995); and idem, *The Politics of Heredity: Essays on Eugenics, Biomedicine, and the Nature-Nurture Debate* (New York: State University of New York Press, 1998).

2   WAR: THE WORLD'S ONLY HYGIENE, 1914–1918

1 Richard A. Soloway, *Demography and Degeneration. Eugenics and the Declining Birthrate in Twentieth-Century Britain* (Chapel Hill: The University of North Carolina Press, 1995), 138.

2 The data is from Michael Howard, *The First World War. A Very Short Introduction* (Oxford: Oxford University Press, 2002), 122. Other excellent discussions of the First World War include Niall Ferguson, *The Pity of War* (London: Allen Lane, 1998); Roland Stromberg, *Redemption by War. The Intellectuals and 1914* (Kansas: The Regents Press of Kansas, 1982); and Ian F. W. Beckett, *The Great War, 1914–18*, 2nd edn. (London: Longman, 2007).

3 F. T. Marinetti, "The Founding and Manifesto of Futurism 1909," in *Futurist Manifestos*, ed. by Umbro Apollonio (London: Thames and Hudson, 1973), 22.

4 Filipo T. Marinetti, "Manifesto agli studenti" in Roger Griffin, ed., *Fascism* (Oxford: Oxford University Press, 1995), 26.

5 Quoted in Roger Griffin, *Fascism* (Oxford: Oxford University Press, 1995), 190.

6 John Alexander Williams, "Ecstasies of the Young: Sexuality, the Youth Movement, and Moral Panic in Germany on the Eve of the First World War," *Central European History* 34, 2 (2001): 163–89.

7 Emilio Gentile, "The Myth of National Regeneration in Italy. From Modernist Avant-Garde to Fascism," in Mattew Affron and Mark Antliff, eds., *Fascist Visions* (Princeton: Princeton University Press, 1997), 38.

8 Kevin Repp, "'More Corporeal, More Concrete': Liberal Humanism, Eugenics, and German Progressives at the Last Fin de Siècle," *The Journal of Modern History* 72, 3 (2000): 683.

9 Karl Pearson, *National Life from the Standpoint of Science* (London: A & C. Black, 1901).

10 J. A. Lindsay, "The Eugenic and Social Influence of the War," *The Eugenics Review* 10, 3 (1918): 133.

11 J. Arthur Thomson, "Eugenics and War," *The Eugenics Review* 6, 1 (1915): 4–5. [italics in the original]

12 J. Arthur Thomson, "Eugenics and War," *The Eugenics Review* 6, 1 (1915): 13.

13 Ibid., 14.

14 J. A. Lindsay, "Eugenics and the Doctrine of the Super-Man," *The Eugenics Review* 6, 3 (1915): 247–62.

15 Gentile, "The Myth of National Regeneration in Italy," 39.

16 Lajos Méhely, *A háború biológiája* (Budapest: Pallas, 1915), 24.

17 Ibid.

18   Theodore G. Chambers, "Eugenics and the War," *The Eugenics Review* 6 (1914/1915): 284.

19   Ibid., 245.

20   Todd Presner, *Muscular Judaism: The Jewish Body and the Politics of Regeneration* (New York: Routledge, 2007), 194–5. See also Maurice Fishberg, "Eugenics in Jewish Life," *The Journal of Heredity* 8, 12 (1917): 543–9.

21   Quoted in Presner, *Muscular Judaism*, 195–6.

22   May Sinclair, *The Tree of Heaven* (London: Cassell, 1917).

23   István Deák, *Beyond Nationalism: A Social and Political History of the Habsburg Officer Corps, 1848–1918* (New York: Oxford University Press, 1990), 195.

24   G. Sergi, "L'eugenica e la Guerra," *Nuova Antologia* 51, 1064 (1916): 135. Quoted in Francesco Cassata, *Molti, sani e forti. L'eugenica in Italia* (Turin: Bollati Boringhieri, 2006), 55.

25   Charles Darwin, *The Descent of Man* (London: Penguin Books, 2004), 160 [first edition 1871].

26   Ploetz, *Grundlinien einer Rassen-Hygiene*, 147.

27   (Anonymous), "Eugenics and War," *The Lancet* 187, 4830 (1916): 685. For similar views, see also Leonard Darwin, "Eugenics During and After the War," *The Eugenics Review* 7, 2 (1915): 91–106.

28   See, for example, Gabriel Petit, Maurice Leudet, eds., *Les Allemands et la science* (Paris: Felix Alcan, 1916).

29   Max von Gruber, *Krieg, Frieden und Biologie* (Berlin: Carl Heymann, 1915), 14.

30   G. Poisson, "La race germanique et sa prétendue supériorité," *Revue anthropologique* 26, 1 (1916): 25–43.

31   L. Capitan, "Les caractères d'infériorité morbide des Austro-Allemands," *Revue anthropologiques* 26, 2 (1916): 75–80. Perhaps the most disturbing feature of this profoundly pathological discourse of racial identity is Capitan's argument that "polychesia" (excessive defecation) and "bromidrosis" (body odor) were distinctive "ethnic" traits. Not surprisingly, then, that in 1917 the Society of Medicine in Paris published a report on the widespread encounter of these diseases in the "German race."

32   Anon., "Eugenics and the War," *The Eugenics Review* 6, 3 (1914): 195–203.

33   F. Savorgnan, *La Guerra e la popolazione* (Bologna: Zanichelli, 1917), 93.

34   Corrado Gini, "The War from the Eugenic Point of View," in *Eugenics in Race and State: Scientific Papers of the Second International Congress of Eugenics*, vol. 2 (Baltimore: Williams & Wilkins, 1923), 430.

35   Leonard Darwin, "On the Statistical Enquiries Needed after the War in Connection with Eugenics," *Journal of the Royal Statistical Society* 7, 2 (1916): 160.

36   Ibid., 161.

37   Ibid., 162.

38   Ibid., 195–6. Emphasis in the original.

39   Ibid., 170.

40 Géza Hoffmann, "Fajegészségtan és népesedési politika," *Természettudományi közlöny* 48, 19/20 (1916): 618.

41 Géza von Hoffmann, *Krieg und Rassenhygiene. Die bevölkerungspolitischen Aufgaben nach dem Kriege* (Munich: J. F. Lehmann, 1916), 7.

42 Ibid., 10–12.

43 Ibid., 16.

44 Ibid.

45 Georg F. Nicolai, *The Biology of War* (London: J. M. Dent & Sons, 1919) [first edition 1917].

46 See J. Tandler, "Bevölkerungspolitische Probleme und Ziele," in *Der Wiederaufbau der Volkskraft nach dem Kriege* (Jena: Verlag von Gustav Fischer, 1918), 95–110.

47 Soloway, *Demography and Degeneration*, 140.

48 Leonard Darwin, "Quality not Quantity," *The Eugenics Review* 7, 4 (1916): 320.

49 Ibid., 321.

50 István Apáthy, "A fajegészségügyi (eugenikai) szakosztály megalakulása," *Magyar Társadalomtudományi Szemle* 7 (1914): 170.

51 See Marius Turda, "The Biology of War: Eugenics in Hungary, 1914–1918," *Austrian History Yearbook* 40, 1 (2009): 238–64.

52 Ladislav Haskovec, "The Eugenics Movement in the Czechoslovak Republic," in *Eugenics in Race and State: Scientific Papers of the Second International Congress of Eugenics*, vol. 2 (Baltimore: Williams & Wilkins, 1923), 440.

53 Lucien March, "Some Attempts towards Race Hygiene in France during the War," *The Eugenics Review* 9, 4 (1918): 198.

54 Géza von Hoffmann, "Eugenics in Germany," *The Journal of Heredity* 5, 10 (1914): 435.

55 Ibid., 345–6.

56 Hermann W. Siemens, *Die biologischen Grundlagen der Rassenhygiene und der Bevölkerungspolitik* (Munich: J. F. Lehmann, 1917).

57 The programme of the Society was published in *Mitteilungen der Deutschen Gesellschaft für Bevölkerungspolitik* 1, 1 (1916): 1–4.

58 Weindling, *Health, Race and German Politics*, 298.

59 Géza von Hoffmann, "Drohende Verflachung und Einseitigkeit rassenhygienischer Bestrebungen in Deutschland," *Archiv für Rassen- und Gesellschaftsbiologie* 12, 3/4 (1917): 343–5.

60 Von Behr-Pinnow, "Zu welchen bevölkerungspolitischen Maßnahmen muß uns der Krieg veranlassen?" *Archiv für Rassen- und Gesellschaftsbiologie* 11, 3 (1915): 333–43.

61 Seth Koven, "Remembering and Dismemberment: Crippled Children, Wounded Soldiers, and the Great War in Great Britain," *The American Historical Review* 99, 4 (1994): 1169.

62 Weiss, *Race Hygiene and National Efficiency*, 141.

63   Neil MacMaster, *Racism in Europe, 1870–2000* (Basingstoke: Palgrave –
      now Palgrave Macmillan, 2001), 46.

64   H. Stöcker, "Staatlicher Gebärzwang oder Rassenhygiene," *Neue
      Generation* 10, 3 (1914): 134–49.

65   János Bársony, "Eugenetik nach dem Kriege," *Archiv für Frauenkunde und
      Eugenetik* 2, 2 (1915): 268.

66   Ibid., 272.

67   Ibid., 275.

68   L. M. Bossi, "Per la difesa della dona e della razza," *La ginecologia moderna*
      10 (1917): 128. Quoted in Cassata, *Molti, sani e forti*, 74. Thanks are due to
      Erin O'Loughlin for the translation.

69   Wilhelm Schallmayer, "Zur Bevölkerungspolitik gegenüber dem
      durch den Krieg verursachten Frauenüberschuß," *Archiv für Rassen- und
      Gesellschaftsbiologie* 11, 6 (1915): 713–37.

70   See Tara Zahra, *Kidnapped Souls: National Indifference and the Battle for
      Children in the Bohemian Lands, 1900–1948* (Ithaca: Cornell University
      Press, 2008); Richard Allen Soloway, *Birth Control and the Population
      Question in England, 1877–1930* (Chapel Hill: The University of North
      Carolina Press, 1982); Edward Ross Dickinson, *The Politics of German Child
      Welfare from the Empire to the Federal Republic* (Cambridge, Mass.: Harvard
      University Press, 1996); Richard Wall, Jay Winter, eds., *The Upheaval of
      War: Family, Work and Welfare in Europe, 1914–1918* (Cambridge: Cambridge
      University Press, 1998).

71   See Ann Taylor Allen, "Feminism and Eugenics in Germany and Britain,
      1900–1940: A Comparative Perspective," *German Studies Review* 23, 3
      (2000): 477–505; and Nancy Wingfield and Maria Bucur, eds., *Gender and
      War in Twentieth Century Eastern Europe* (Bloomington: Indiana University
      Press, 2006).

72   Kristen Stromberg Childers, *Fathers, Families, and the State, 1914–1945*
      (Ithaca: Cornell University Press, 2003), 3.

73   Richard Allen Soloway, *Birth Control and the Population Question in England,
      1877–1930* (Chapel Hill: University of North Carolina Press, 1982), 173.

74   In 1916, a conference was organised in Berlin dealing with the prob-
      lem of demography, eugenics and population policy. See *Die Erhaltung
      und Vermehrung der deutschen Volkskraft. Verhandlungen der 8. Konferenz
      der Zentralstelle für Volkswohlfahrt in Berlin vom 26. bis 28. Oktober 1915*
      (Berlin: Heymann, 1916). In Hungary, a conference on national hygiene
      was organised in 1918. See Béla Fenyvessy, József Madzsar, eds., *A
      népegészségi országos nagygyűlés munkálatai* (Budapest: Eggenberger,
      1918). There were, however, conferences aiming at an international
      consensus, such was the conference on preventive medicine organ-
      ised with German, Austrian, Hungarian and Bulgarian doctors in
      Berlin in February 1918 and the Austro-Hungarian conference on

child rearing and protection organised in 1918 in Vienna. The first German and Austro-Hungarian conference on racial hygiene and population policy, although announced for September 1918 did not take place.

75 See Marius Turda, "The First Debates on Eugenics in Hungary, 1910–1918," in Turda, Weindling, eds., *Blood and Homeland*, 185–221.

76 Géza von Hoffmann, "Eugenics in the Central Empires since 1914," *Social Hygiene* 7, 3 (1921): 291.

77 Ibid., 291–4.

78 István Apáthy, "A fajegészségtan köre és feladatai," *Természettudományi Közlöny* 50, 691/2, part 2 (1918): 86.

79 Géza von Hoffmann, "Rassenhygiene in Ungarn," *Archiv für Rassen- und Gesellschaftsbiologie*, 14, 1 (1918): 55–67.

80 See, for example, A. L. Galéot, *L'avenir de la race: Le problème du peuplement en France* (Paris: Nouvelle librairie nationale, 1917).

81 Modris Ekstein, *Rites of Spring: The Great War and the Birth of the Modern Age* (London: Macmillan, 2000), 325. [first edition 1989]

82 Theodore Szél, "The Genetic Effects of War in Hungary," in *Scientific Papers of the Third International Congress of Eugenics* (Baltimore: Williams & Wilkins, 1934), 251.

83 Ibid., 252.

84 Ibid., 254.

85 George L. Mosse, *Fallen Soldiers: Reshaping the Memory of the World Wars* (Oxford: Oxford University Press, 1990), 7.

3 EUGENIC TECHNOLOGIES OF NATIONAL IMPROVEMENT, 1918–1933

1 Gentile, "The Myth of National Regeneration in Italy," 26.

2 Marius Turda, "The Nation as Object: Race, Blood and Biopolitics in Interwar Romania," *Slavic Review* 66, 3 (2007): 413–41.

3 One illustrative example is Alfred Ploetz, "Die rassenbiologische Bedeutung des Krieges und sein Einfluß auf den deutschen Menschen," *Volk und Rasse* 6, 3 (1931): 148–55.

4 Illustrative in this sense is the series of conferences organised by the French Eugenics Society at l'Ecole des Hautes-Etudes in 1920 and 1921. The general theme was the consequences of war from a eugenic point of view and guest speakers included Eugène Apert, Lucien March, Charles Richet, Georges Schreiber and Leonard Darwin. The lectures were then published as *Eugénique et selection* (Paris: Felix Alcan, 1922).

5 Ladislav Haskovec, "The Eugenics Movement in the Czechoslovak Republic," 436.

6 August Forel, *The Sexual Question. A Scientific, Psychological, Hygienic and Sociological Study for the Cultured Classes* (New York: Rebman, 1908), 511.

7 Ibid., 512.

8 Erving Goffman, *Stigma. Notes on the Management of Spoiled Identity* (New York: Simon & Schuster, 1986), 4 [first edition 1963].

9 Galton, "Hereditary Improvement," 129.

10 Ibid., 5. For a more recent perspective see Todd F. Heatherton, Robert E. Kleck, Michelle R. Hebl, and Jay G. Hull, eds., *The Social Psychology of Stigma* (New York: Guilford Press, 2000).

11 Quoted in Marjatta Hietala, "From Race Hygiene to Sterilization: The Eugenics Movement in Finland," in Gunnar Broberg, and Nils Roll-Hansen, eds., *Eugenics and the Welfare State: Sterilization Policy in Denmark, Sweden, Norway, and Finland* (East Lansing: Michigan State University Press, 2005), 220–1.

12 Ignaz Kaup, "Was kosten die minderwertigen Elemente dem Staat und der Gesellschaft?" *Archiv für Rassen- und Gesellschaftsbiologie* 10, 12 (1913): 747.

13 Quoted in Koch, "Eugenic Sterilization in Scandinavia," 302.

14 Andrija Štampar, "On Health Politics," *Jugoslavenska njiva* 29–31 (1919): 1–29. Reproduced in *Serving the Cause of Public Health. Selected Papers of Andrija Štampar*, ed. by M. D. Grmek (Zagreb: Andrija Štampar School of Public Health, 1966), 62.

15 Illustrative in this sense is the appeal of the French anthropologists to the wider academic community of the Entente powers to strengthen their collaboration in the fields of anthropology and eugenics and to counter "German imperialism." See "Appel aux anthropologists alliés," *Revue anthropologique* 29, 1/2 (1919): 52–4.

16 H. Laughlin, "National Eugenics in Germany," *The Eugenics Review* 12, 4 (1920/21): 304–7.

17 Due to its provocative assertions, Baur's text was not translated and published in English as Baur intended. It was published for the first time by Bentley Glass in "A Hidden Chapter of German Eugenics between the Two World Wars," *Proceedings of the American Philosophical Society* 125, 5 (1981): 363–5.

18 Ibid., 364.

19 Marshall Berman, *All that Is Solid Melts into Air: The Experience of Modernity* (New York: Penguin, 1988), 15 [first edition 1982].

20 Vladislav Růžička, "A Motion for the Organization of Eugenical Research," in *Scientific Papers of the Second International Congress of Eugenics.* vol. 2, 452.

21 Hermann W. Siemens, *Race Hygiene and Heredity* (New York: London, 1924), 127 [first edition 1917].

22 As the author of the much celebrated *Die Rassenhygiene in den Vereinigten Staaten von Nordamerika* (Munich: J. F. Lehmann, 1913), Hoffmann had published extensively on American eugenics before 1918.

23 Géza von Hoffmann, "New Eugenics in Hungary," *The Journal of Heredity* 11, 1 (1920): 41.

24 Raymond Pearl, "Sterilization of Degenerates and Criminals considered from the Standpoint of Genetics," *The Eugenics Review* 10, 1 (1919): 6.

25 See Paul Weindling, "International Eugenics: Swedish Sterilization in Context," *Scandinavian Journal of History* 24, 2 (1999): 179–97.

26 Weindling, *Health, Race and German Politics*, 307.

27 Quoted in David G. Horn, *Social Bodies. Science, Reproduction, and Italian Modernity* (Princeton: Princeton University Press, 1994), 57.

28 Gisela Bock, "Racism and Sexism in Nazi Germany: Motherhood, Compulsory Sterilization, and the State," *Signs: Journal of Women in Culture and Society* 8, 3 (1983): 404. See also her recent "Nationalsozialistische Sterilisationpolitik," in Klaus-Dietmar Henke, ed., *Tödliche Medizin im Nationalsozialismus. Von der Rassenhygiene zum Massenmord* (Cologne: Böhlau, 2008), 85–99.

29 John Macnicol, "Eugenics and the Campaign for Voluntary Sterilization in Britain between the Wars," *Social History of Medicine* 2, 2 (1989): 155–6.

30 Quoted in Michal Šimůnek, "Eugenics, Social Genetics and Racial Hygiene: Plans for the Scientific Regulation of Human Heredity in the Czech Lands, 1900–1925," in Turda, Weindling, eds., *Blood and Homeland*, 153.

31 See Michelle Mouton, *From Nurturing the Nation to Purifying the Volk. Weimar and Nazi Family Policy, 1918–1945* (New York: Cambridge University Press, 2007); Atina Grossmann, *Reforming Sex: The German Movement for Birth Control and Abortion Reform, 1920–1950* (Oxford: Oxford University Press, 1995); Claudia Koonz, *Mothers in the Fatherland: Women, the Family and Nazi Politics* (New York: St Martin's Press, 1987); Renate Bridenthal, Atina Grossmann, and Marion Kaplan, eds., *When Biology became Destiny: Women in Weimar and Nazi Germany* (New York: Monthly Review Press, 1984).

32 See Paul J. Weindling's "A City Regenerated: Eugenics, Race and Welfare in Interwar Vienna," in Deborah Holmes and Lisa Silverman, eds., *Interwar Vienna: Culture between Tradition and Modernity* (New York: Camden House, 2009), 81–113.

33 Quoted in Hjalmar Anderson, "The Swedish State-Institute for Race-Biological Investigations: An Account of its Origination," in *The Swedish Nation in Word and Picture, together with short summaries of the contribution made by Swedes within the fields of Anthropology, Race-Biology, Genetics and Eugenics* (Stockholm: Hasse W. Tullberg, 1921), 52–3.

34 Růžička, "A Motion for the Organization of Eugenical Research," 452.

35 Quoted in Rory Yeomans, "Of 'Yugoslav Barbarians' and Croatian Gentlemen Scholars: Nationalist Ideology and Racial Anthropology in Interwar Yugoslavia," in Turda, Weindling, eds., *Blood and Homeland*, 90.

36 Quoted in ibid., 92.

37 Karl Pearson, "Editorial," *Annals of Eugenics* 1, 1–2 (1925): 3–4.

38 Iuliu Moldovan, *Igiena națiunii: Eugenia* (Cluj: Institutul de Igienă și Igienă Socială, 1925). See also Maria Bucur, *Eugenics and Modernization in Interwar Romania* (Pittsburgh: Pittsburg University Press, 2002).

39 Moldovan, *Igiena*, 46.

40 Otto Reche, *Die Bedeutung der Rassenpflege für die Zukunft unseres Volkes* (Vienna: Wiener Gesellschaft für Rassenpflege, 1925). See also Katja Geisenhainer, *Rasse ist Schicksal: Otto Reche (1879–1966): Ein Leben als Anthropologe und Volkerkundler* (Leipzig: Evangelische Verlagsanstalt, 2002).

41 Károly Balás, *The Foundation of Social Politics* (Ilford: C. W. Clark, 1926), 9.

42 Stavros Zurukzoglu, *Biologische Probleme der Rassenhygiene und der Kulturvölker* (Munich: J.F. Bergmann, 1925), 170–4.

43 Alfred Grotjahn, *Die Hygiene der menschlichen Forpflanzung. Versuch einer Praktischen Eugenik* (Berlin: Urban & Schwarzenberg, 1926), vii.

44 H. Lundborg, "Race Biological Institutes," in *Proceedings of the World Population Conference*, ed. by Margaret Sanger (London: Edward Arnold, 1927), 350.

45 Quoted in Hjalmar Anderson, "The Swedish State-Institute for Race-Biological Investigation," 49.

46 Iuliu Hațieganu, "Rolul social al medicului în opera de consolidare a statului național," *Transilvania* 54 (1925): 588.

47 Ibid., 590.

48 "The Meeting of the International Federation of Eugenic Organizations," *Eugenical News* 14, 11 (1929): 154.

49 Drieu La Rochelle, *Notes pour comprendre le siècle* (Paris: Gallimard, 1941), 153.

50 Karl Binding, Alfred Hoche, *Die Freigabe der Vernichtung Lebensunwertem Lebens. Ihr Mass und Ihre Form* (Leipzig: Felix Meiner, 1920).

51 Karl Binding and Alfred Hoche, "Permitting the Destruction of Unworthy Life," in *The Nazi Germany Sourcebook. An Anthology of Texts*, ed. by Roderick Stackelberg, Sally A. Winkle (New York: Routledge, 2002), 71.

52 Ibid., 71–2.

53 Ibid., 72.

54 Ibid., 73.

55 Julius Tandler, *Gefahren der Minderwertigkeit* (Vienna: Verlag des Wiener Jugendhilfswerks, 1929).

56 Edmond Székely, *Sexual Harmony and the New Eugenics* (London: C. W. Daniel Comp., 1938).

57 Manuel Devaldès, *La maternité consciente. Le role des femmes dans l'amélioration de la race* (Paris: Radot, 1927), 62.

58 Eugen Relgis, *Umanitarism şi eugenism* (Bucharest: 'Vegetarismul', 1935), 28–29 (emphasis in the original).

59 Madison Grant, *The Passing of the Great Race, or the Racial Basis of European History* (New York: Charles Scribner's Sons, 1916), 49.

60 From "Leitsatze der Deutschen Gesellschaft für Rassen-hygiene," translated by Paul Popenoe in "Eugenics in Germany," *The Journal of Heredity* 13, 8 (1922): 382–4.

61 Gunnar Broberg, Nils Roll-Hansen, eds., *Eugenics and the Welfare State. Sterilization Policy in Denmark, Sweden, Norway, and Finland* (East Lansing: Michigan State University, 2005) [first edition 2007].

62 H. Lundborg, *Degenerationsfaran och riktlinjer för dess förebyggande* (Stockholm, 1922). Quoted in Broberg, Nils Roll-Hansen, eds., *Eugenics and the Welfare State*, 85.

63 Ioan Manliu, *Crâmpeie de eugenie şi igienă socială* (Bucharest: Tip. 'Jockey-Club', 1921), 21.

64 Ibid.

65 *Nachrichten der Wiener Gesellschaft für Rassenpflege* 1 (1924): 9.

66 Maria Bucur, *Eugenics and Modernization in Interwar Romania* (Pittsburgh: Pittsburgh University Press, 2002), 8.

67 St. Zurukzoglu, "Die Probleme der Eugenik unter Besonderer Berücksichtigung der Verhütung Erbkranken Nachwuchses," in St. Zurukzoglu, ed., *Verhütung Erbkranken Nachwuchses. Eine kritische Betrachtung und Würdigung* (Basel: Benno Schwabe, 1938), 7–57.

68 Harry W. Paul, "Religion and Darwinism. Varieties of Catholic Reactions," in Thomas F. Glick, ed., *The Comparative Reception of Darwinism* (Chicago: Chicago University Press, 1988), 404.

69 "The Legalisation of Eugenic Sterilisation," *The Lancet* 219, 2 (1930): 360.

70 For the stipulations of the law see Zurukzoglu, ed., *Verhütung Erbkranken Nachwuchses*, 264–6. See also Philippe Ehrenström, "Eugenisme et santé publique: la stérilisation légale des malades mentaux dans le canton de Vaud (Suisse)," *History and Philosophy of the Life Sciences*, 15, 2 (1993): 205–27; and Volker Roelcke, "Zeitgeist und Erbgesundheitsgesetzgebung im Europe der 1930er Jahre. Eugenik, Genetik und Politik im historischen Kontext," *Der Nervenarzt* 73, 11 (2002): 1019–30.

71 "Danish Sterilization Law," *Eugenical News* 14, 8 (1929): 122; and Zurukzoglu, ed., *Verhütung Erbkranken Nachwuchses*, 281–7. The 1929 Law was reformed in 1934 and 1935, respectively.

72 See the discussion in Mathew Thomson, *The Problem of Mental Deficiency. Eugenics, Demography, and Social Policy in Britain, c. 1870–1959* (Oxford: Oxford University Press, 1998), 180–206.

73 Cora B. S. Hodson, *Human Sterilization Today: A Survey of Current Practice* (London: Watts, 1934).

74 Ernst Rüdin, "The Significance of Eugenics and Genetics for Mental Hygiene," in *Proceedings of the First International Congress on Mental Hygiene*. vol. 1 (New York: The International Committee for Mental Hygiene, 1932), 471–88.

75 Ladislaus Benedek, "Eugenical Efforts in Hungary," *Eugenical News* 16, 10 (1931): 173.

76 See Turda, *A Healthy Nation*.

77 Quoted in Sevasti Trubeta, "Eugenic Birth Control and Prenuptial Health Certificates in Interwar Greece," in Christian Promitzer, Sevasti Trubeta, Marius Turda, eds., *Hygiene, Health and Eugenics in Southeastern Europe to 1945* (Budapest: Central European University Press, 2010).

78 Quoted in Gergana Mircheva, "Marital Health and Eugenics in Bulgaria, 1878–1940," in ibid.

79 Liviu Stan, *Rasă şi religiune* (Sibiu: Tiparul Tipografie Arhidiecezane, 1942).

80 Quoted in Griffin, ed., *Fascism*, 191.

81 Quoted in Mircheva, "Marital Health and Eugenics."

82 Ioan Manliu, "Sterilizarea degeneraţilor," *Revista de igienă socială*, 1, 5 (1931): 382.

83 "Ibid., 382–3.

84 Quoted in Maria Falina, "Between 'Clerical Fascism' and Political Orthodoxy: Orthodox Christianity and Nationalism in Interwar Serbia," in Matthew Feldman, Marius Turda, eds., *Clerical Fascism in Interwar Europe* (London: Routledge, 2008), 40.

85 Sabine Schleiermacher, *Sozialethik im Spannungsfeld von Sozial- und Rassenhygiene: Der Mediziner Hans Harmsen im Centralausschuss für die Innere Mission* (Husum: Matthiesen, 1998).

86 Quoted in Young-Sun Hong, *Welfare, Modernity, and the Weimar State, 1919–1933* (Princeton: Princeton University Press, 1998), 10.

87 Hans Harmsen, "Wie entwickelt sich die Eheberatung," *Die Innere Mission* 23 (1928): 249.

88 Quoted in Paul A. Hanebrink, *In Defense of Christian Hungary. Religion, Nationalism, and Antisemitism, 1890–1944* (Ithaca: Cornell University Press, 2006), 132.

89 Tudor Georgescu, "The Eugenic Fortress: Alfred Csallner and the Saxon Eugenic Discourse in Interwar Romania," in Promitzer, Trubeta, Turda, eds., *Hygiene, Health and Eugenics in Southeastern Europe to 1945* (2010).

90 Paul, "Religion and Darwinism," 404.

91 Etienne Lepicard, "Eugenics and Roman Catholicism: An Encyclical Letter in Context: *Casti Connubii*, December 31, 1930," *Science in Context* 11, 3/4 (1998): 533.

92 Joseph Mayer, "Eugenics in Roman Catholic Literature," *Eugenics* 3, 2 (1930): 43–51.

93 Quoted in William H. Schneider, *Quality and Quantity: The Quest for Biological Regeneration in Twentieth-Century France* (Cambridge: Cambridge University Press, 1990), 192.

94 Jean Dermine, "Les lois du mariage et les devoir des époux," and Mgr Dubourg, "Le véritable eugénisme" in *L'Église et l'eugénisme. La famille a la croisée des chemins* (Paris: Éditions Mariage et Famille, 1930), 174–90 and 224–7.

95 Johann Ude, *Der moralische Schwachsinn. Für Volkssittlichkeit* (Graz: Eigenverlag, 1918).

96 Ingrid Richter, *Katholizismus und Eugenik in der Weimarer Republic und im Dritten Reich: Zwischen Sittlichkeitsreform und Rassenhygiene* (Paderborn: Ferdinand Schöningh, 2001).

97 Monika Löscher, "Eugenic and Catholicism in Interwar Austria," in Turda, Weindling, eds., *Blood and Homeland*, 303.

98 See, for example, Merman Muckermann, *Eugenik und Katholizismus* (Berlin: Alfred Metzner, 1933); Albert Niedermeyer, "Die Sterilisierung vor dem Forum der Wissenschaft und der Moral," *St. Lukas* 4, 3 (1936): 97–120; and Tihamér Tóth, *Eugenik vom katholischen Standpunkt* (Vienna: Raimund Fürlinger, 1937).

99 "Casti Connubii, Encyclical of Pope Pius XI on Christian Marriage to the Venerable Brethren, Patriarchs, Primates, Archbishops, Bishops, and other local Ordinaries enjoying Peace and Communion with the Apostolic See," in *The Eugenics Movement: An International Perspective*, ed. by Pauline M. H. Mazumdar, vol. 4 (New York: Routledge, 2007), 81–2.

100 Ibid.

101 Ibid., 81–2.

102 Hermann Muckermann, "Eugenics and Catholicism," in *The Eugenics Movement: An International Perspective*, ed. by Pauline M. H. Mazumdar, vol. 4 (New York: Routledge, 2007), 22.

103 Ibid., 67.

4   EUGENICS AND BIOPOLITICS, 1933–1940

1 The literature on these subjects is so extensive that even to attempt a selection of the most important works is difficult. However, for a recent discussion of Nazi family policies see Mouton, *From Nurturing the Nation to Purifying the Volk*; for the impact of Nazi racial policies on

German society, see Richard J. Evans, *The Third Reich in Power, 1933–1939* (London: Allen Lane, 2005) and Götz Aly, *Hitler's Beneficiaries: Plunder, Racial War, and the Nazi Welfare State* (London: Macmillan, 2007).

2 On Nazi racial research in Eastern Europe, Michael Burleigh's *Germany turns Eastwards: A Study of Ostforschung in the Third Reich* (Cambridge: Cambridge University Press, 1988) remains the standard reading. On Nazi racial anthropology see Christopher M. Hutton, *Race and the Third Reich* (Cambridge: Polity Press, 2005). On recent interpretations of the Holocaust see Dan Stone, ed., *The Historiography of the Holocaust* (Basingstoke: Palgrave Macmillan, 2004).

3 Quoted in Ilinca Zarifopol-Johnston, *Searching for Cioran* (Bloomington: Indiana University Press, 2009), 243.

4 Griffin, *Modernism and Fascism*, 253.

5 Peter Fritzsche, *Germans into Nazis* (Cambridge, Mass.: Harvard University Press, 1998).

6 R. Walther Darré, *Neuadel aus Blut und Boden* (Munich: J. F. Lehmann, 1930), 190.

7 Quoted in Griffin, ed., *Fascism*, 136.

8 Fritz Lenz, "The Position of National Socialism on Race Hygiene," in *The Eugenics Movement*, vol. 4, 19. See also Fritz Lenz, "Zur Frage eines Sterilisierungsgesetzes," *Eugenik. Erblehre, Erbpflege* 3, 4 (1933): 73–6.

9 Arthur Gütt, Ernst Rüdin, Falk Ruttke, *Gezetz zur Verhütung erbkranken Nachwuchses vom 14. Juli 1933 mit Auszug aus dem Gesetz gegen gefährliche Gewohnheitsverbrecher und über Massregeln der Sicherung und Besserung vom 24. Nov. 1933* (Munich: J. F. Lehmann, 1934). See the translation of the law in Paul Popenoe: "The German Sterilisation Law," *Journal of Heredity*, 25, 7 (1934): 257–9 and "Eugenical Sterilization in Germany," *Eugenical News* 18, 5 (1933): 91–3. For a discussion of the law and its implication for the racial revolution of the Nazi state, see Michael Burleigh, Wolfgang Wippermann, *The Racial State: Germany 1933–1945* (Cambridge: Cambridge University Press, 1991).

10 Marie Kopp, "Legal and Medical Aspects of Eugenic Sterilization in Germany," *American Sociological Review* 1, 5 (1936): 770.

11 Walter Gross, "Grundfragen nationalsozialistischer Rassen und Bevölkerungspolitik," (1941). Quoted in Griffin, ed. *Fascism*, 157.

12 See, for example, Martin Staemmler, "Die Sterilisierung Minderwertiger vom Standpunkt des Nationalsozialismus," *Eugenik. Erblehre, Erbpflege* 3, 5 (1933): 97–110.

13 Hilda von Hellmer Wullen, "Eugenics in Other Lands: A Survey of Recent Developments," *The Journal of Heredity* 28, 8 (1937): 269.

14 Gisela Bock, "Sterilization and 'Medical' Massacres in National Socialist Germany: Ethic, Politics, and the Law," in Manfred Berg and Geoffrey Cocks, eds., *Medicine and Modernity: Public Health and Medical*

Care in Nineteenth-and Twentieth Century Germany (Cambridge: Cambridge University Press, 1997), 150.

15 As eloquently illustrated by the contributors to Ernst Rüdin, ed., *Erblehre und Rassenhygiene im völksichen Staat* (Munich: J. F. Lehmanns, 1934).

16 C. Thomalia, "The Sterilization Law in Germany," *Eugenical News* 19, 6 (1934): 140–1.

17 Wullen, "Eugenics in Other Lands," 275.

18 "Voluntary Sterilization Bill," *The Eugenics Review* 27, 2 (1935): 136.

19 Hans Maier, "On Practical Experience of Sterilization in Switzerland," *The Eugenics Review* 26, 1 (1934): 24.

20 B. Sekla, "A Czech Eugenicist, Professor Dr Vladislav Růžička, and the Czechoslovak Eugenics Society that He Founded," *Eugenical News* 20, 6 (1935): 102.

21 Wullen, "Eugenics in Other Lands," 271.

22 "Sterilization in Hungary," *Eugenical News* 19, 6 (1934): 142.

23 Nils von Hofsten, "Sterilization in Sweden," *The Eugenics Review* 29, 4 (1938): 257; and Zurukzoglu, ed., *Verhütung Erbkranken Nachwuchses*, 299.

24 "Sterilization Bill for Norway," *Eugenical News* 18, 5 (1933): 94–5. Swedish, Finish and Estonian sterilisation laws are included in Zurukzoglu, ed., *Verhütung Erbkranken Nachwuchses*, 300–6. See also Nils Roll-Hansen, "Norwegian Eugenics: Sterilization as Social Reform," and Marjatta Hietala, "From Race Hygiene to Sterilization," in Broberg, Gunnar, Roll-Hansen, eds., *Eugenics and the Welfare State*, 151–94 and 195–258, respectively.

25 Quoted in Gheorghe Marinescu, "Despre hereditatea normală şi patologică şi raporturile ei cu eugenia," *Memoriile Secţiunii Ştiinţifice* 3, 11 (1936): 70.

26 Ibid., 71.

27 See Ayça Alemdaroğlu, "Politics of the Body and Eugenic Discourse in Early Republican Turkey," *Body & Society* 11, 3 (2004): 86–101; and Murat Ergin, "Biometrics and Anthropometrics: The Twins of Turkish Modernity," *Patterns of Prejudice* 42, 3 (2008): 281–304.

28 Naci Somersan, "Prenuptial Medical Examination in Turkey," *The Eugenics Review* 29, 4 (1938): 261–3.

29 Theophil Laanes, "Eugenics in Estonia," *Eugenical News* 20, 6 (1935): 103.

30 In 1932, the General Secretary of the French Eugenics Society, Henri Vignes, invited several physicians and sociologists, including E. Apert and G. Schreiber, to express their opinions on the issue of sterilization, publishing the debate as "Stérilisation des inadaptés sociaux," *Revue anthropologique* 42, 7/9 (1932): 228–44.

31 Georges Schreiber, "Actual Aspect of the Problem of Eugenical Sterilization in France," *Eugenical News* 11, 5 (1936): 105. See also Georges Schreiber, "La Stérilisation Eugénique en Allemagne," *Revue anthropologique* 45, 1/3 (1935): 84–91.

32 *Premier Congrès Latin d'Eugénique. Rapport* (Paris: Masson, 1937), 6.

33 G. Banu, "Les facteurs dysgéniques en Roumanie. Principes d'un Programme Pratique d'Eugenique," in *Premier Congrès Latin d'Eugénique. Rapport* (Paris: Masson, 1937), 319.

34 G. Banu, *L'hygiène de la race. Étude de biologie héréditaire et de normalisation de la race* (Paris: Masson, 1939), 297.

35 See Marius Turda, "'To End the Degeneration of a Nation," 98–100.

36 Jon Alfred Mjöen, "Further Directions for a Race Hygienic Population-Politic," *Eugenical News* 11, 5 (1936): 114.

37 Gheorghe Banu, "Critical and Synthetical Examination of the Rural Health Problems," in *Problemele sanitare ale populaţiei rurale din România*, vol. 10, part 2 (Bucharest: Revista de igienă socială, 1940), 1407.

38 Quoted in Aaron Gillette, *Racial Theories in Fascist Italy* (London: Routledge, 2002), 39.

39 Quoted in ibid., 40.

40 Corneliu Zelea Codreanu, *For My Legionaries* (York, SC.: Liberty Bell Publications, 2003), 312.

41 Ibid., 313.

42 Ibid., 315.

43 L. v. Méhely, "Blut und Rasse," *Zeitschrift für Morphologie und Anthropologie* 34 (1934): 257.

44 Nichifor Crainic, "George Cosbuc, Poetul rasei noastre," in Nichifor Crainic, *Puncte cardinale în haos* (Bucharest: Albatros, 1998), 120–1 [first edition 1936].

45 Leone Lattes, *Individuality of the Blood in Biology and in Clinical and Forensic Medicine* (London: Oxford University Press, 1932), 43 [first edition 1923].

46 "Manifesto of Racial Scientists" (1938), translated by Aaron Gillette, "The Origins of the 'Manifesto of Racial Scientists'," *Journal of Modern Italian Studies* 6, 3 (2001): 319.

47 Lazër Radi, *Fashizmi dhe fryma shqiptare* (Tirana: Distapur, 1940) 87. Translation provided by Rigels Halili.

48 Ibid., 153.

49 Quoted in Marina Petrakis, *The Metaxas Myth: Dictatorship and Propaganda in Greece* (London: I. B. Tauris, 2006), 126.

50 Štefan Polakovič, "Slovak National Socialism" (1940), in Marius Turda, Diana Mishkova, eds., *Anti-Modernism: Radical Revisions of Collective Identity* (Budapest: Central European University Press, forthcoming 2011).

51 *Quinsling ruft Norwegen! Reden und Aufsätze* (Munich: Franz Eher, 1942), 136.

52 Emanuel Vajtauer, "Czech Mythos" (1943), in Turda, Mishkova, eds., *Anti-Modernism* (2011).

53 See Marius Turda, "From Craniology to Serology: Racial Anthropology in Interwar Hungary and Romania," *Journal of the History of Behavioral Sciences* 43, 3 (2007), 361–77.

54 John Koumaris, "On the Morphological Variety of Modern Greeks," *Man* 48, 141 (1948): 126.

55 Ibid., 127.

56 Ibid., 127. Although this text was published in 1948, Koumaris' ideas of the "fluid constancy" of the race were already formulated in the late 1930s. See Trubeta, "Anthropological Discourses and Eugenics in Interwar Greece," 135.

57 Quoted in Alemdaroğlu, "Politics of the Body," 73.

58 Alemdaroğlu, "Politics of the Body," 69. See also Ayça Alemdaroğlu, "Eugenics, Modernity and Nationalism," in David Turner, Kevin Stagg, eds., *Social Histories of Disability and Deformity* (London: Routledge, 2006), 126–41.

59 Iordache Făcăoaru, "Normele eugenice în organizaţiile legionare," *Cuvântul* 17, 69 (1940): 1

60 Corrado Gini, "Das Bevölkerungsproblem Italiens und die fascistische Bevölkerungspolitik," *Archiv für Rassen- und Gesellschaftsbiologie* 25, 1 (1931): 18.

61 Quoted in Artemis Leontis, *Topographies of Hellenism: Mapping the Homeland* (Ithaca: Cornell University Press, 1995), 114.

62 Leontis, *Topographies of Hellenism*, p. 115.

63 P. P. Panaitescu, "Noi suntem de aici," *Cuvântul* 17, 38 (1940): 1.

64 Iordache Făcăoaru, *Structura rasială a populaţiei rurale din România* (Bucharest: F. Göbl, 1940), 16 (italics in the original).

65 Ken Kalling, "The Self-Perception of a Small-Nation: The Reception of Eugenics in Interwar Estonia," in Turda, Weindling, eds., *Blood and Homeland*, 253.

66 Harry Federley, "Rassenhygienische Propagandaarbeit unter der schwedischen Bevölkerung Finnlands," *Archiv für Rassen- und Gesesellschaftsbiologie* 24 (1930): 236. Euthenics concentrated mostly on hygienic education, improving living conditions and preventive medicine. See Lester F. Ward, "Eugenics, Euthenics, and Eudemics," *The American Journal of Sociology* 18, 6 (1913): 737–54.

67 Quoted in Ossian Schauman, "Eugenic Work in Swedish Finland," *The Swedish Nation in Word and Picture*, 94.

68 "Heimatbildung und Volksgestaltung," *Sudetendeutsches Jahrbuch* 1 (1924): 115–19.

69 See, for example, Konrad Henlein, "Die völkische Sendung der Frau," in *Konrad Henlein Spricht. Reden zur politischen Volksbewegung der Sudetendeutschen* (Karlsbad: Verlag Karl H. Frank, 1937), 90–100.

70 One should mention here the *German Society for Family Research and Eugenics in the Czechoslovak Republic* established in 1938.

71 Alfred Csallner, *Das Landesamt für Statistik und Sippenwesen der Deutschen Volksgemeinschaft in Rumänien* (1939), 1. Quoted in Georgescu, "The Eugenics Fortress," in Promitzer, Trubeta, Turda, eds., *Hygiene, Health and Eugenics in Southeastern Europe to 1945* (2010).

72 The role played by the German communities of Central and Eastern Europe in the Nazi prospects for a new racial order in the East has been subject to much research. See, for example, Balász A. Szelényi, "From Minority to Übermensch: The Social Roots of Ethnic Conflict in the German Diaspora of Hungary, Romania and Slovakia," *Past and Present* 196, 1 (2007): 215–51; Valdis O. Lumans, *Himmler's Auxiliaries: The Volksdeutsche Mittelstelle and the German National Minorities of Europe, 1933–1945* (Chapel Hill: University of North Carolina Press, 1993); and Anthony Komjáthy, Rebecca Stockwell, *German Minorities and the Third Reich: Ethnic Germans of East Central Europe between the Wars* (New York: Holms and Meier, 1980).

73 See Christof Morrissey, "Ethnic Politics and Scholarly Legitimation: The German Institut für Heimatforschung in Slovakia, 1941–1944," in Ingo Haar, Michael Fahlbusch, eds., *German Scholars and Ethnic Cleansing, 1919–1945* (New York: Berghahn Books, 2005), 100–9.

74 Petru Râmneanţu, *Die Abstammung der Tschangos* (Sibiu: Centrul de studii şi cercetări cu privire la Transylvania, 1944).

75 Quoted in Cassata, *Molti, sani e forti*, 223.

76 Quoted in ibid., 227. See also Francesco Cassata, *"La Difesa della razza". Politica, ideologia e immagine del razzismo fascista* (Torino: Giulio Einaudi, 2008).

77 Quoted in *Martiriul evreilor din Romania, 1940–1944. Documente şi mărturii* (Bucharest: Hasefer, 1991), 16.

78 Charles Balás, *Malthus and the Population Problems of Today* (Budapest: Stephaneum, 1936), 31.

79 "German Population and Race Politics. An Address by Dr Frick, Reichminister for the Interior, before the First Meeting of the Expert Council for Population- and Race-Politics in Berlin, June 28, 1933," *Eugenical News* 19, 2 (1934): 36.

80 Ibid., 38.

81 Quoted in "A French View. A Study of National Policies which Purpose to Influence Eugenical Trends along Definitely Pre-Determined Lines," *Eugenical News* 19, 2 (1934): 39.

82 G. W. Harris, "Bio-Politics," *The New Age* 10, 9 (1911): 197.

83 See, for example, Morley Roberts, *Bio-Politics: An Essay in the Physiology, Pathology and Politics of the Social and Somatic Organism* (London: Dent, 1938). For a recent discussion see Roberto Esposito, *Bios: Biopolitics and Philosophy* (Minneapolis: University of Minnesota Press, 2008).

84 Edward Ross Dickinson, "Biopolitics, Fascism, Democracy: Some Reflections on Our Discourse about 'Modernity'," *Central European History* 37, 1 (2004): 1.

85 Ibid., 42.

86 Iuliu Moldovan, *Biopolitica* (Cluj: Institutul de igienă şi igienă socială, 1926), 5.

87 Ibid., 17. For a general discussion of Moldovan's ideas, see Bucur, *Eugenics and Modernization*, 78–121.

88 Eugen Fischer, *Der völkische Staat, biologisch gesehen* (Berling: Junker und Dünnhaupt, 1933), 5.

89 Ibid., 22. For similar opinions see Martin Staemmler, *Rassenpflege im völkischen Staat* (Munich: J. F. Lehmann, 1933) and Walter Schultze, "Die Bedeutung der Rassenhygiene für Staat und Volk in Gegenwart und Zukunft," in Rüdin, ed., *Erblehre und Rassenhygiene*, 1–21.

90 Quoted in Maria Sophia Quine, *Italy's Social Revolution: Charity and Welfare from Liberalism to Fascism* (Basingstoke: Palgrave – now Palgrave Macmillan, 2002), 135–6.

91 Lajos Antal, *A biologizmus mint új életszemlélet. A magyar biopolitika* (Budapest: Magyar Egyetemi Nymoda, 1940), 8.

92 Traian Herseni, "Mitul sângelui," *Cuvântul* 17, 41 (1940): 2.

93 Traian Herseni, "Rasă şi destin naţional," *Cuvântul* 18, 91 (1941): 1.

94 Ibid., 7.

95 Michel Foucault, *"Society Must be Defended". Lectures at the Collège de France, 1975–1976* (New York: Picador, 1997), 242. See also Michel Foucault, *The History of Sexuality: An Introduction* (New York: Vintage Books, 1980), 136–40.

96 Mihai Antonescu, "Directive şi îndrumări date inspectorilor administrativi şi pretorilor trimişi în Basarabia şi Bucovina," in *Martiriul evreilor din România, 1940–1944. Documente şi mărturii* (Bucharest: Hasefer, 1991), 139.

97 Although the ethnographer Ion Chelcea was suggesting as late as 1944 that certain Roma groups should be "settled in an isolated region, transferred to Transnistria and, if necessary, sterilised." See Ion Chelcea, *Ţiganii din România. Monografie etnografică* (Bucharest: Editura Institutului Central de Statistică, 1944), 101.

98 Zygmunt Bauman, *Modernity and the Holocaust* (Cambridge: Polity Press, 1989).

99 Galton, "Probability, The Foundation of Eugenics," 99.

CONCLUSION: TOWARDS AN EPISTEMOLOGY OF EUGENIC KNOWLEDGE

1 Carl E. Schorske, *Thinking with Modernity: Explorations in the Passage to Modernism* (Princeton: Princeton University Press, 1998), 230.

2 Peter Bowler, *The Mendelian Revolution: The Emergences of Hereditarian Concepts in Modern Science and Society* (London: The Athlone Press, 1989), 17. Bowler expands here on the "strong programme in the sociology of knowledge" proposed by David Boor in his *Knowledge and Social Imagery* (Chicago: Chicago University Press, 1991) [first edition 1976].

3 Gabrielle M. Spiegel, "The Task of the Historian," *American Historical Review* 114, 1 (2009): 2.

4 Zygmunt Bauman, *Modernity and Ambivalence* (London: Polity Press, 1991); Michael Schwartz, "Biopolitik in der Moderne", *Internationale wissenschaftliche Korrespondenz der deutschen Arbeiterbewegung*, vol. 31 (1995); Tzvetan Todorov, *Hope and Memory: Lessons from the Twentieth Century* (Princeton: Princeton University Press, 2004); Roger Griffin, *Modernism and Fascism: The Sense of a New Beginning under Mussolini and Hitler* (Basingstoke: Palgrave Macmillan, 2007); and Aristotle Kallis, *Genocide and Fascism: The Eliminationist Drive in Fascist Europe* (London: Routledge, 2009).

5 Stepan, *"The Hour of Eugenics"*, 6.

6 See, from a long list of studies, Jürgen Habermas, *The Future of Human Nature* (London: Polity Press, 2003); Nicholas Agar, *Liberal Eugenics: In Defence of Human Enhancement* (Oxford: Blackwell, 2004); Jean Gayon, Daniel Jacobi, eds., *L'éternal retour de l'eugénisme* (Paris: Presses Universitaires de France, 2006); and John Harris, *Enhancing Evolution: The Ethical Case for Making Better People* (Princeton: Princeton University Press, 2007).

7 Darwin, *The Descent of Man*, 689.

8 Galton, *Inquiry into Human Faculty*, 220.

SELECT BIBLIOGRAPHY

I have organised this bibliography into two major sections, one deal-ing with primary sources, the other with secondary literature. As the bibliography on eugenics is exceptionally diverse and vast, those interested in further readings as well as more in-depth and country-specific analysis can rely on additional bibliographies provided by most of the works mentioned in the secondary literature.

PRIMARY SOURCES

Anderson, Hjalmar. "The Swedish State-Institute for Race-Biological Investigations: An Account of its Origination," in *The Swedish Nation in Word and Picture, together with short summaries of the contribution made by Swedes within the fields of Anthropology, Race-Biology, Genetics and Eugenics.* Ed. by H. Lundborg and J. Runnström. Stockholm: Hasse W. Tullberg, 1921. pp. 48–56.

Antal, Lajos. *A biologismus mint új életszemlélet. A magyar biopolitika.* Budapest: Magyar Egyetemi Nyomda, 1940.

Apáthy, István. "A fajegészségügyi (eugenikai) szakosztály megalakulása," *Magyar Társadalomtudományi Szemle* 7 (1914): 165–72.

——. "A fajegészségtan köre és feladatai," *Természettudományi Közlöny* 50, 689/92, 2 parts (1918): 6–21 and 81–101.

Balás, Károly. *The Foundation of Social Politics.* Ilford: C. W. Clark, 1926.

——. *Malthus and the Population Problems of Today.* Budapest: Stephaneum, 1936.

Banu. Gheorghe. "Les facteurs dysgéniques en Roumanie. Principes d'un Programme Pratique d'Eugenique," in *Premier Congrès Latin d'Eugénique. Rapport.* Paris: Masson, 1937. pp. 296–319.

——. *L'Hygiene de la race. Étude de biologie héréditaire et de normalisation de la race.* Paris: Masson, 1939.

——. "Critical and Synthetical Examination of the Rural Health Problems," in *Problemele sanitare ale populației rurale din România,* vol. 10, part 2. Bucharest: Revista de igienă socială, 1940. pp. 1399–515.

Bársony, János. "Eugenetik nach dem Kriege," *Archiv für Frauenkunde und Eugenetik* 2, 2 (1915): 267–75.

Baur, Erwin. "Eugenics in the New Germany," in Bentley Glass, "A Hidden Chapter of German Eugenics between the Two World Wars," *Proceedings of the American Philosophical Society* 125, 5 (1981): 363–5.

Behr-Pinnow, Carl von. "Zu welchen bevölkerungspolitischen Maßnahmen muß uns der Krieg veranlassen?" *Archiv für Rassen- und Gesellschaftsbiologie* 11, 3 (1915): 333–43.

Bell, Alexander G. *A Few Thoughts Concerning Eugenics*. Washington: Press of Judd & Detweiler, 1908.

——. "How to Improve the Race," *Journal of Heredity* 5, 1 (1914): 1–7.

Benedek, Ladislaus. "Eugenical Efforts in Hungary," *Eugenical News* 16, 10 (1931): 172–3.

Binding, Karl and Alfred Hoche, *Die Freigabe der Vernichtung lebensunwertem Lebens: ihr Mass und ihre Ziel*. Leipzig: Felix Meiner, 1920.

Bloomfield, Paul. "The Eugenics of the Utopians: The Utopia of the Eugenists," *The Eugenics Review* 40, 4 (1949): 191–8.

Boldrini, Marcello. "Some Dysgenical Effects of the War in Italy," *Social Hygiene* 7, 3 (1921): 265–78.

Capitan, L. "Les caractères d'inferiorité morbide des Austro-Allemands," *Revue anthropologiques* 26, 2 (1916): 75–80.

Carrel, Alexis. *Man, the Unknown*. London: Hamish Hamilton, 1935.

"Casti Connubii, Encyclical of Pope Pius XI on Christian Marriage to the Venerable Brethren, Patriarchs, Primates, Archbishops, Bishops, and other local Ordinaries enjoying Peace and Communion with the Apostolic See," in *The Eugenics Movement: An International Perspective*. Ed. by Pauline M. H. Mazumdar. vol. 4. New York: Routledge, 2007. pp. 69–100.

Chambers, Theodore G. "Eugenics and the War," *The Eugenics Review* 6, 4 (1914): 271–90.

Chelcea, Ion. *Ţiganii din România. Monografie etnografică*. Bucharest: Editura Institutului Central de Statistică, 1944,

Codreanu, Corneliu Zelea. *For My Legionaries*. York, SC.: Liberty Bell Publications, 2003 [first edition 1936].

Constantinescu, Gheorghe. *Ereditate şi eugenie*. Bucharest: Torouţiu, 1936.

Coulter, John M. "What Biology Has Contributed to Religion," *The Biblical World* 41, 4 (1913): 219–23.

Crainic, Nichifor. *Puncte cardinale in haos*. Bucharest: Albatros, 1998 [first edition 1936].

Csallner, Alfred. *Die volksbiologische Forschung unter den Siebenbürger Sachsen und ihre Auswirkung auf das Leben diese Volksgruppe*. Leipzig: S. Hirzel, 1940.

Darwin, Charles. *The Descent of Man*. London: Penguin Books, 2004 [first edition 1871].

Darwin, Leonard. "Heredity and Environment. A Warning to Eugenicists," *The Eugenics Review* 7, 2 (1916): 93–122.

Darwin, Leonard. "Quality not Quantity," *The Eugenics Review* 7, 4 (1916): 297–321.

——. "On the Statistical Enquiries after the War in Connection with Eugenics," *Journal of the Royal Statistical Society* 79, 2 (1916): 159–88.

——. "The Need for Widespread Eugenic Reform during Reconstruction," *The Eugenics Review* 9, 3 (1918): 145–62.

——. "The Aims and Methods of Eugenical Societies," *Science* 54, 1397 (1921): 313–23.

——. "The Field of Eugenic Reform," *The Scientific Monthly* 13, 5 (1921): 385–98.

——. *What is Eugenics?* London: Watts, 1929.

Dermine, Jean. "Les lois du mariage et les devoir des époux," in *L'Église et l'eugénisme. La famille a la croisée des chemins*. Paris: Éditions Mariage et Famille, 1930. pp. 174–90.

Devaldès, Manuel *La Maternite Consciente. Le Role des Femmes dans l'Amélioration de la Race*. Paris: Radot, 1927.

Dubourg, Mrsg. "Le véritable eugénisme," in *L'Église et l'eugénisme. La famille a la croisée des chemins*. Paris: Éditions Mariage et Famille, 1930. pp. 224–7.

Ellis, Havelock. *The Problem of Race-Regeneration*. London: Cassell, 1911.

——. "Eugenics during and after the War," *The Eugenics Review* 7, 2 (1915): 91–106.

Făcăoaru, Iordache. "Normele eugenice în organizațiile legionare," *Cuvântul* 17, 69 (21 December 1940): 1–2.

——. *Structura rasială a populației rurale din România*. Bucharest: F. Göbl, 1940.

Federley, Harry. "Rassenhygienische Propagandaarbeit under der schwedischen Bevölkerung Finnlands," *Archiv für Rassen- und Gesellschaftsbiologie* 24 (1930): 326–33.

Field, James A. "The Progress of Eugenics," *The Quarterly Journal of Economics* 26, 1 (1911): 1–67.

——. "Problems of Population after the War," *The American Economic Review* 7, 1 (1917): 233–7.

Fischer, Eugen. *Der völkische Staat, biologisch gesehen*. Berlin: Junker und Dünnhaupt, 1933.

——. *Le problème de la race et la législation raciale en Allemagne*. Paris: Fernand Sorlot, 1942.

Fishberg, Maurice. "Eugenics in Jewish Life," *The Journal of Heredity* 8, 12 (1917): 543–49.

Forel, August. "Die Alkoholfrage als Cultur- und Rassenproblem," in *Bericht über den VIII. Internationalen Congress gegen den Alkoholismus*. Ed. by Rudolf Wlassak. Vienna: F. Deuticke, 1902. pp. 29–34.

——. *The Sexual Question. A Scientific, Psychological, Hygienic and Sociological Study for the Cultured Classes*. New York: Rebman, 1908.

——. *Malthusianism oder Eugenik?* Munich: Verlag von Ernst Reinhardt, 1911.

Foustka, Břetislav. *Slabí v lidské společnosti. Ideály humanitní a degenerace národů*. Prague: Jana Laichtera, 1904.

Foustka, Břetislav. *Die Abstinenz als Kulturproblem mit besonderer Berücksichtigung der österreichischen Völkerstämme*. Vienna: Verlag von Brüder Suschitzky, 1908.

Galéot, A. L. *L'avenir de la race: Le problème du peuplement en France*. Paris: Nouvelle librairie nationale, 1917.

Galton, Francis. "Hereditary Character and Talent," *MacMillan's Magazine* 12, 68 (1865): 157–66 and 12, 70 (1865): 318–27.

——. *Hereditary Genius: An Inquiry into its Laws and Consequences*. London: Macmillan, 1869.

——. "Hereditary Improvement," *Fraser's Magazine* 7, 37 (1873): 116–30.

——. *Inquiry into Human Faculty and Its Development*. London: Macmillan, 1883.

——. "Presidential Address," in *Transactions of the Seventh International Congress of Hygiene and Demography*, vol. 10. London: Eyre and Spottiswoode, 1892. pp. 7–12.

——. "Eugenics: Its Definition, Scope, and Aims," *The American Journal of Sociology* 10, 1 (1904): 1–25.

——. "Studies in Eugenics," *The American Journal of Sociology* 11, 1 (1905): 11–25.

——. *Memories of My Life*, 2nd edn. London: Methuen, 1908.

——. *Essays in Eugenics*. London: Eugenics Education Society, 1909.

Gasset, José Ortega. "History as a System," in Raymond Klibansky and H. J. Paton, eds., *Philosophy and History*. London: Oxford University Press, 1936. pp. 283–322.

Gerrard, Thomas J. *The Church and Eugenics*. London: P.S. King & Son, 1912.

Gini, Corrrado. "The War from the Eugenic Point of View," in *Scientific Papers of the Second International Congress of Eugenics*. vol. 2. Baltimore: Williams & Wilkins, 1923. pp. 430–31.

——. "The Scientific Basis of Fascism," *Political Science Quarterly* 42, 1 (1927): 99–115.

——. "Das Bevölkerungsproblem Italiens und die fascistische Bevölkerungspolitik," *Archiv für Rassen- und Gesellschaftsbiologie* 25, 1 (1931): 1–18.

——. "Report of the Committee for the Study of the Eugenic and Dysgenic Effects of War," in *Scientific Papers of the Third International Congress of Eugenics*. Baltimore: Williams & Wilkins, 1934. pp. 231–43.

Goldstein, Kurt. *Über Rassenhygiene*. Berlin: Julius Springer, 1913.

Gordon, Ernest. *The Anti-Alcohol Movement in Europe*. New York: Fleming H. Revell, 1913.

Grant, Madison. *The Passing of the Great Race, or the Racial Basis of European History*. New York: Charles Scribner's Sons, 1916.

Grotjahn, Alfred. *Soziale Pathologie: Versuch einer Lehre von den sozialen Beziehungen der menschlichen Krankheiten als Grudlage der soziale Medizin und der sozialen Hygiene*. Berlin: Hirschwald, 1912.

Grotjahn, Alfred. *Die Hygiene der menschlichen Fortpflanzung. Versuch einer praktischen Eugenik*. Berlin: Urban & Schwarzenberg, 1926.

Gruber, Max von. *Krieg, Frieden und Biologie*. Berlin: Carl Heymann, 1915.

Gütt, Arthur and Ernst Rüdin, Falk Ruttke. *Gezetz zur Verhütung erbkranken Nachwuchses vom 14. Juli 1933 mit Auszug aus dem Gesetz gegen gefährliche Gewohnheitsverbrecher und über Massregeln der Sicherung und Besserung vom 24. Nov. 1933*. Munich: J. F. Lehmann, 1934.

Harris, G. W. "Bio-Politics," *The New Age* 10, 9 (1911): 197.

Haškovec, Ladislav. "Moderne eugenische Bewegung," *Wiener Klinische Rundschau* 26, 39 (1912): 609–11 (part I and II); 26, 40 (1912): 625–7 (part III); 26, 40 (1912): 643–5 (part IV) and 26, 42 (1912): 659–61 (final part).

——. "The Eugenics Movement in the Czechoslovak Republic," in *Scientific Papers of the Second International Congress of Eugenics*. vol. 2. Baltimore: Williams & Wilkins, 1923. pp. 435–42.

Hațieganu, Iuliu. "Rolul social al medicului în opera de consolidare a statului național," *Transilvania* 54 (1925): 587–91.

Hayek, F. A. von. "Scientism and the Study of Society," *Economica* 9, 35 (1942): 267–91.

Hellmer, Hilda von. "Eugenics in Other Lands: A Survey of Recent Developments," *The Journal of Heredity* 28, 8 (1937): 269–75.

Hericourt, J. *Les maladies de la société*. Paris: Flammarion, 1918.

Herseni, Traian. "Mitul sângelui", *Cuvântul* 17, 41 (1940): 1–2.

——. "Rasă și destin național," *Cuvântul* 18, 91 (1941): 1 and 7.

Hirschfeld L. and H. Hirschfeld. "Serological Differences between the Blood of Different Races," *The Lancet* 197, 2 (1919): 675–9.

Hodson, Cora B. S. *Human Sterilization Today: A Survey of Current Practice*. London: Watts, 1934.

——. "Eugenics in Norway," *The Eugenics Review* 27, 1 (1935): 41–4.

——. "International Federation of Eugenic Organizations. Report of the 1936 Conference," *The Eugenics Review* 28, 3 (1936): 217–19.

Hoffmann, Géza. *Die Rassenhygiene in den Vereinigten Staaten von Nordamerika*. Munich: J. F. Lehmann, 1913.

——. "Ausschüsse für Rassenhygiene in Ungarn," *Archiv für Rassen- und Gesellschafsbiologie* 10, 6 (1914): 830–1.

——. "Eugenics in Germany," *The Journal of Heredity* 5, 10 (1914): 435–6.

——. "Drohende Verflachung und Einseitigkeit rassenhygienischer Bestrebungen in Deutschland," *Archiv für Rassen- und Gesellschaftsbiologie* 12, 3/4 (1916): 343–5.

——. "Fajegészségtan és népesedési politika," *Természettudományi közlöny* 48, 19/20 (1916): 617–21.

——. *Krieg und Rassenhygiene. Die bevölkerungspolitischen Aufgaben nach dem Kriege*. Munich: J. F. Lehmanns, 1916.

Hoffmann, Géza. "Rassenhygiene in Ungarn," *Archiv für Rassen- und Gesellschaftsbiologie*, 13, 1 (1918): 55–67.

——. "New Eugenics in Hungary," *The Journal of Heredity* 11, 1 (1920): 41.

——. "Eugenics in the Central Empires since 1914," *Social Hygiene* 7, 3 (1921): 285–96.

Hofsten, Hils von. "Sterilization in Sweden," *The Eugenics Review* 29, 4 (1938): 257–60.

Jacoby, Gerhard. *Racial State. The German Nationalities Policies in the Protectorate of Bohemia-Moravia*. New York: Institute of Jewish Affairs, 1944.

Jahn, Rudolf. *Konrad Henlein: Leben und Werk des Turnführers*. Karlsbad: Adam Kraft Verlag, 1938.

Johnson, Roswell H. "Eugenics and so-called Eugenics," *The American Journal of Sociology* 20, 1 (1914): 98–103.

——. *International Eugenics*. PhD Dissertation. University of Pittsburgh, 1934.

Jojkić, Mladen. *Pokušaj fiziološko-patološke studije o srpskom narodu*. Subotica: Štamparija Vinka Blesića, 1895.

Jordan, David Starr. "The Eugenics of War," *The Eugenics Review* 5, 3 (1914): 197–213.

Jordan, David Starr, and Harvey Ernest Jordan. *War's Aftermath: A Preliminary Study of the Eugenics of War as Illustrated by the Civil War of the United States and the late Wars in the Balkans*. Boston: New York: Houghton Mifflin, 1914.

Kaup, Ignaz. "Was kosten die minderwertigen Elemente dem Staat und der Gesellschaft?" *Archiv für Rassen- und Gesellschaftsbiologie* 10, 12 (1913): 723–47.

Kleinsmid, Rufus Bernhard von. *Eugenics and the State*. Jeffersonville: Indiana Reformatory Printing Trade School, 1913.

Konrad Henlein Spricht. Reden zur politischen Volksbewegung der Sudetendeutschen. Karlsbad: Verlag Karl H. Frank, 1937.

Kopp, Marie E. "Legal and Medical Aspects of Eugenic Sterilization in Germany," *American Sociological Review* 1, 5 (1936): 761–70.

——. "Eugenic Sterilization Laws in Europe," *American Journal of Obstetrics and Gynecology* 34, 3 (1937): 499–504.

Koumaris, John. "On the Morphological Variety of Modern Greeks," *Man* 48, 141 (1948): 126–7.

La Rochelle, Drieu. *Notes pour comprendre le siècle*. Paris: Gallimard, 1941.

Laanes, Theophil. "Eugenics in Estonia," *Eugenical News* 20, 6 (1935): 103–4.

Landau, E. "Notes Eugéniques," *Revue anthropologiques* 26, 7/8 (1916): 310–13.

Lankester, Ray E. *Degeneration: A Chapter in Darwinism*. London: Macmillan, 1880.

Lattes, Leone. *Individuality of the Blood in Biology and in Clinical and Forensic Medicine*. London: Oxford University Press, 1932 [first edition 1923].

Laughlin, Harry H. *The Legal Status of Eugenical Sterilization*. Chicago: Municipal Court of Chicago, 1929.

Lenz, Fritz. "Zum Begriff der Rassenhygiene und seiner Benennung," *Archiv für Rassen- und Gesellschaftsbiologie* 11, 4 (1915): 445–8.

———. "Eugenics in Germany," *The Journal of Heredity* 15, 5 (1924): 223–31.

———. "Zur Frage eines Sterilisierungsgesetzes," *Eugenik. Erblehre, Erbpflege* 3, 4 (1933): 73–6.

———. "The Position of National Socialism on Race Hygiene," in *The Eugenics Movement: An International Perspective*. Ed. by Pauline M. H. Mazumdar. vol. 4. New York: Routledge, 2007. pp. 13–19.

Lindsay, J. A. "Eugenics and the Doctrine of the Super-Man," *The Eugenics Review* 6, 3 (1915): 247–62.

———. "The Eugenic and Social Influence of the War," *The Eugenics Review* 9, 3 (1918): 133–44.

Lundborg, H. "Race Biological Institutes," in *Proceedings of the World Population Conference*. Ed. by Margaret Sanger. London: Edward Arnold, 1927. pp. 47–50.

———. *Bevölkerungsfragen, Bauerntum und Rassenhygiene*. Berlin: Alfred Metzner, 1934.

Maier, Hans. "On Practical Experience of Sterilization in Switzerland," *The Eugenics Review* 26, 1 (1934): 19–25.

Manliu, Ioan. *Crâmpeie de eugenie şi igienă socială*. Bucharest: Tip. 'Jockey-Club', 1921.

———. "Sterilizarea degeneraţilor," *Revista de igienă socială* 1, 5 (1931): 374–85.

March, Lucien. "Depopulation and Eugenics," *The Eugenics Review* 5, 3 (1914): 234–51 and 5, 4 (1914): 343–51.

———. "Some Attempts towards Race Hygiene in France during the War," *Eugenics Review* 9, 4 (1918): 195–212.

———. "The Consequence of War and the Birth Rate in France," *The Scientific Monthly* 13, 5 (1921): 399–419.

Marinescu, Gheorghe. "Despre hereditatea normală şi patologică şi raporturile ei cu eugenia," *Memoriile Secţiunii Ştiinţifice* 3, 11 (1936): 1–85.

Matiegka, J. "Ueber den Einfluss des Alkohols auf die geistigen und moralischen Eigenschaften der Bevölkerung Böhmens," *Bericht über den VIII. Internationalen Congress gegen den Alkoholismus*. Ed. by Rudolf Wlassak. Vienna: F. Deuticke, 1902. pp. 339–53.

Mayer, Joseph. "Eugenics in Roman Catholic Literature," *Eugenics* 3, 2 (1930): 43–51.

Méhely, Lajos. *A háború biológiája*. Budapest: Pallas, 1915.

———. "Blut und Rasse," *Zeitschrift für Morphologie und Anthropologie* 34 (1934): 244–57.

Metchnikoff, Elie. *The Nature of Man. Studies in Optimistic Philosophy*. London: G. P. Putman's Sons, 1903.

Mjöen, Jon Alfred. "Rassenhygiene in Norwegen," *Archiv für Rassen- und Gesellschaftsbiologie* 24 (1930): 334–41.

——. "Further Directions for a Race Hygienic Population-Politic," *Eugenical News* 11, 5 (1936): 112–14.

Moldovan, Iuliu. *Igiena naţiunii: Eugenia*. Cluj: Institutul de Igienă şi Igienă Socială, 1925.

——. *Biopolitica*. Cluj: Institutul de Igienă şi Igienă Socială, 1926.

More, Adelyne. *Fecundity versus Civilisation*. London: George Allen & Unwin, 1917 [first edition 1916].

Muckermann, Hermann. *Eugenik*. Berlin: Dümmler, 1934.

——. *Eugenik und Katholizismus*. Berlin: Alfred Metzner, 1933.

——. "Eugenics and Catholicism," in *The Eugenics Movement: An International Perspective*. Ed. by Pauline M. H. Mazumdar. vol. 4. New York: Routledge, 2007. pp. 21–67.

Mügge, Maximilian A. *Eugenics and the Superman. A Racial Science and a Racial Religion*. London: Eugenics Education Society, 1909.

Naeser, Vincent. "Eugenics Marriage Bills in the Scandinavian Countries," *The Eugenics Review* 5, 3 (1914): 238–9.

Nicolai, Friedrich G. *The Biology of War*. London: J. M. Dent & Sons, 1919 [first edition 1917].

Niedermeyer, Albert. "Die Sterilisierung vor dem Forum der Wissenschaft und der Moral," *St. Lukas* 4, 3 (1936): 97–120.

Nisot, Marie-Thérèse. "La sterilisation des anormaux," *Mercure de France* 209, 375 (1929): 595–603.

Nordau, Max. *Degeneration*. Lincoln: University of Nebraska Press, 1993 [first edition 1892].

Onslow H. "The French Commission on Depopulation," *The Eugenics Review* 5 (1913): 130–52.

Osborn, Frederick. "Development of a Eugenic Philosophy," *American Sociological Review* 2, 3 (1937): 389–97.

Osborn, Henry Fairfield. "Eugenics: The American and Norwegian Programs," *Science* 54, 1403 (1921): 482–4.

Oswald, Frances. "Eugenical Sterilization in the United States," *The American Journal of Sociology* 36, 1 (1930): 65–73.

Panaitescu, P. P. "Noi suntem de aici," *Cuvântul* 17, 38 (20 November 1940): 1.

Pearl, Raymond. "Sterilization of Degenerates and Criminals considered from the Standpoint of Genetics," *The Eugenics Review* 10, 1 (1919): 1–6.

Pearson, Karl. *National Life from the Standpoint of Science*. London: A & C. Black, 1901.

——. *The Function of Science in the Modern State*. Cambridge: Cambridge University Press, 1902.

——. *The Groundwork of Eugenics*. London: Dulau, 1909.

——. *The Problem of Practical Eugenics*. London: Dulau, 1909.

Pearson, Karl. "Editorial," *Annals of Eugenics* 1, 1/2 (1925): 3–4.

Petit, Gabriel, Maurice Leudet, eds., *Les Allemands et la science*. Paris: Felix Alcan, 1916.

Petrini-Galatzi, Mihail. *Filosofia medicală: Despre amelioraţiunea rasei umane.* Bucharest: Tipografia D. A. Laurian, 1876.

Ploetz, Alfred. *Grundlinien einer Rassen-Hygiene. vol. 1. Die Tüchtigkeit unserer Rasse und der Schutz der Schwachen.* Berlin: S. Fischer, 1895.

——. "Die Begriffe Rasse und Gesellschaft und die davon abgeleiteten Disciplinen," *Archiv für Rassen- und Gesellschaftsbiologie* 1, 1 (1904): 1–27.

——. "Der Alkohol im Lebensprozeß der Rasse," in *Bericht über den IX. Internationalen Kongress gegen den Alkoholismus (Bremen 14–19.IV.1903).* Ed. by Franziskus Hähnel. Jena: Verlag von Gustav Fischer, 1904. pp. 70–95.

——. "Neo-Malthusianism and Race Hygiene," in *Problems in Eugenics. Report of the Proceedings of the First International Eugenics Congress*, vol. 2. London: The Eugenics Education Society, 1913. pp. 183–9.

——. "Die rassenbiologische Bedeutung des Krieges und sein Einfluß auf den deutschen Menschen," *Volk und Rasse* 6, 3 (1931): 148–55.

Poisson, G. "La race germanique et sa prétendue supériorité," *Revue anthropologique* 26, 1 (1916): 25–43.

Popenoe, Paul. "Eugenics in Germany," *The Journal of Heredity* 13, 8 (1922): 382–4.

——. "The German Sterilization Law," *The Journal of Heredity* 25, 7 (1934): 257–60.

Popenoe, Paul and Roswell Hill Johnson. *Applied Eugenics.* New York: Macmillan, 1920.

Quinsling ruft Norwegen! Reden und Aufsätze. Munich: Franz Eher, 1942.

Radi, Lazër. *Fashizmi dhe fryma shqiptare.* Tirana: Distapur, 1940.

Râmneanţu, Petru. *Die Abstammung der Tschangos.* Sibiu: Centrul de studii şi cercetări cu privire la Transilvania, 1944.

Relgis, Eugen. *Umanitarism şi eugenism.* Bucharest: "Vegetarismul", 1935.

Rentoul, Robert R. *Race Culture; Or, Race Suicide?* London: Walter Scott Publishing, 1906.

Rice, Thurman B. *Racial Hygiene. A Practical Discussion of Eugenics and Race Culture.* New York: Macmillan, 1929.

Roberts, Morley. *Bio-Politics: An Essay in the Physiology, Pathology and Politics of the Social and Somatic Organism.* London: Dent, 1938.

Rüdin, Ernst. "Der Alkohol im Lebensprozeß der Rasse," *Bericht über den IX. Internationalen Kongress gegen den Alkoholismus (Bremen 14–19.IV.1903).* Ed. by Franziskus Hähnel. Jena: Verlag von Gustav Fischer, 1904. pp. 95–107.

——. "The Significance of Eugenics and Genetics for Mental Hygiene," in *Proceedings of the First International Congress on Mental Hygiene.* vol. 1. Ed. by Frankwood E. Williams. New York: The International Committee for Mental Hygiene, 1932. pp. 471–88.

Rüdin, Ernst. ed. *Erblehre und Rassenhygiene im völksichen Staat*. Munich: J. F. Lehmanns, 1934.

Růžička. Vladislav. "A Motion for the Organization of Eugenical Research," in *Scientific Papers of the Second International Congress of Eugenics*. vol. 2. Baltimore: Williams & Wilkins, 1923. pp. 452–5.

——. *Biologické základy eugeniky*. Prague: Fr. Borový, 1923.

Saleeby, Caleb Williams. *Heredity*. London: T. C. & E. C. Jack, 1905.

——. *Parenthood and Race Culture: An Outline of Eugenics*. London: Cassell, 1909.

——. *The Methods of Race-Regeneration*. London: Cassell, 1911.

Savorgnan, Franco. *La Guerra e la popolazione*. Bologna: Zanichelli, 1917.

——. "L'influence de la guerre sur le mouvement naturel de la population," *Scientia* 25, 34/5 (1919): 381–91.

Schallmayer, Wilhelm. *Über die drohende physische Entartung der Culturvölker* Berlin: Heuser, 1895.

——. "Zur Bevölkerungspolitik gegenüber dem durch den Krieg verursachten Frauenüberschuß," *Archiv für Rassen- und Gesellschaftsbiologie* 11, 6 (1915): 713–37.

Schauman, Ossian. "Eugenic Work in Swedish Finland," in *The Swedish Nation in Word and Picture, together with short summaries of the contribution made by Swedes within the fields of Anthropology, Race-Biology, Genetics and Eugenics*. Ed. by H. Lundborg and J. Runnström. Stockholm: Hasse W. Tullberg, 1921. pp. 89–94.

Schreiber, Georges. "La Stérilisation Eugénique en Allemagne," *Revue anthropologique* 45, 1–3 (1935): 84–91.

——. "Actual Aspect of the Problem of Eugenical Sterilization in France," *Eugenical News* 11, 5 (1936): 104–5.

Schuster, Edgar. *Eugenics*. London: Collins, 1912.

Schweisheimer, W. "Bevölkerungsbiologie Bilanz des Krieges 1914/19," *Archiv für Rassen- und Gesellschaftsbiologie* 13, 2/4 (1920): 176–93.

Schultze, Walter. "Die Bedeutung der Rassenhygiene für Staat und Volk in Gegenwart und Zukunft," in Ernst Rüdin, ed., *Erblehre und Rassenhygiene im völkischen Staat*. Munich: J. F. Lehmann, 1934. pp. 1–21.

Schuman, Frederick. "The Political Theory of German Fascism," *The American Political Science Review*, 28, 2 (1934): 210–32.

Sekla, B. "Czech Eugenicist, Professor Dr Vladislav Růžička, and the Czechoslovak Eugenics Society that He Founded," *Eugenical News* 20, 6 (1935): 101–3.

——. "Eugenics in Czechoslovakia," *The Eugenics Review* 28, 2 (1936): 115–7.

Sergi, Giuseppe. *Le Degenerazioni Umane*. Milano: Fratelli Dumolard, 1889.

——. *La Decadenza della Nazione Latine*. Milano: Fratelli Bocca, 1900.

Siebert, Friedrich. *Der völkische Gehalt der Rassenhygiene*. Munich: J. F. Lehmann, 1917.

Siegmund, Heinrich, *Zur sächsischen Rassenhygiene*. Hermannstadt: Peter Drotleff, 1901.

Siemens, Hermann W. "Biologische Terminologie und Rassenhygienische Propaganda," *Archiv für Rassen- und Gesellschaftsbiologie* 12, 3–4 (1917): 257–67.

——. *Die biologischen Grundlagen der Rassenhygiene und der Bevölkerungspolitik*. Munich: J. F. Lehmanns Verlag, 1917.

——. *Race Hygiene and Heredity*. New York: D. Appleton, 1924.

Sinclair, May. *The Tree of Heaven*. London: Cassell, 1917.

Somersan, Naci. "Prenuptial Medical Examination in Turkey," *The Eugenics Review* 29, 4 (1938): 261–3.

Staemmler, Martin. *Rassenpflege im völkischen Staat*. Munich: J. F. Lehmann, 1933.

——. "Die Sterilisierung Minderwertiger vom Standpunkt des Nationalsozialismus," *Eugenik. Erblehre, Erbpflege* 3, 5 (1933): 97–110.

Stan, Liviu. *Rasă și religiune*. Sibiu: Tiparul Tipografie Arhidiecezane, 1942.

Stöcker, Helene. "Staatlicher Gebärzwang oder Rassenhygiene," *Neue Generation* 10, 3 (1914): 134–49.

Székely, Edmond. *Sexual Harmony and the New Eugenics*, trans. and ed. by Purcell Weaver. London: C. W. Daniel Comp., 1938.

Szél, Theodore. "The Genetic Effects of the War in Hungary," in *Scientific Papers of the Third International Congress of Eugenics*. Baltimore: Williams & Wilkins, 1934. pp. 249–54.

S. [Alexandru Sutzu], "Evoluțiunea și hereditatea," *Gazeta medico-chirurgicală a spitalelor* 5, 12 (1874): 182–7.

Tandler, J. "Bevölkerungspolitische Probleme und Ziele," in *Der Wiederaufbau der Volkskraft nach dem Kriege*. Jena: Verlag von Gustav Fischer, 1918. pp. 95–110.

——. *Gefahren der Minderwertigkeit*. Vienna: Verlag des Wiener Jugendhilfswerks, 1929.

Thomalia, C. "The Sterilization Law in Germany," *Eugenical News* 19, 6 (1934): 137–42.

Thomson, Arthur J. "Eugenics and War," *The Eugenics Review* 6, 1 (1915): 1–14.

Tóth, Tihamér. *Eugenik vom katholischen Standpunkt*. Vienna: Raimund Fürlinger, 1937.

Ude, Johann. *Der moralische Schwachsinn. Für Volkssittlichkeit*. Graz: Eigenverlag, 1918.

Vignes, Henri. "Stérilisation des inadaptés sociaux," *Revue anthropologique* 42, 7–9 (1932): 228–44.

Voegelin, Eric. "The Origins of Scientism," *Social Research* 15, 4 (1948): 462–494.

Ward, Lester F. "Eugenics, Euthenics, and Eudemics," *The American Journal of Sociology* 18, 6 (1913): 737–54.

Whetham, William C. D. and Catherine D. Whetham. *An Introduction to Eugenics.* London: Macmillan, 1912.

Zurukzoglu, Stavros. *Biologische Probleme der Rassenhygiene und der Kulturvölker.* Munich: J. F. Bergmann, 1925.

———. ed. *Verhütung Erbkranken Nachwuchses. Eine kritische Betrachtung und Würdigung.* Basel: Benno Schwabe, 1938.

———. "Die Probleme der Eugenik unter Besonderer Berücksichtigung der Verhütung Erbkranken Nachwuchses," in *Verhütung Erbkranken Nachwuchses. Eine kritische Betrachtung und Würdigung.* Ed. by St. Zurukzoglu. Basel: Benno Schwabe, 1938. pp. 7–57.

* * *

"A fajnemesítés (eugenika) problémái," *Huszadik Század* 23, 12 (1911): 29–44; 157–70; 322–36 and 694–709.

"A French View. A Study of National Policies which Purpose to Influence Eugenical Trends along Definitely Pre-Determined Lines," *Eugenical News* 19, 2 (1934): 39–40.

"Appel aux anthropologists alliés," *Revue anthropologique* 29, 1–2 (1919): 52–4.

"Comptes Rendus de la Société Hellénique d'Anthropologie," *Revue anthropologique* 38, 7–9 (1928): 291–8.

"Danish Sterilization Law," *Eugenical News* 14, 8 (1929): 122–4.

"Eugenics in Austria," *The Eugenics Review* 26, 4 (1935): 259–61.

"Eugenical Sterilization in Germany," *Eugenical News* 18, 5 (1933): 89–93.

"German Population and Race Politics. An Address by Dr Frick, Reichminister for the Interior, before the First Meeting of the Expert Council for Population- and Race-Politics in Berlin, June 28, 1933," *Eugenical News* 19, 2 (1934): 33–8.

"Heimatbildung und Volksgestaltung," *Sudetendeutsches Jahrbuch* 1 (1924): 115–9.

"International Eugenics Congress," *Science* 53, 1358 (1921): 16–17.

"Jewish Eugenics," *The Journal of Heredity* 8, 2 (1917): 72–4.

"Sterilization Bill for Norway," *Eugenical News* 18, 5 (1933): 94–5.

"Sterilization in Hungary," *Eugenical News* 19, 6 (1934): 142.

"The Legalisation of Eugenic Sterilisation," *The Lancet* 219, 2 (1930): 360.

"The Meeting of the International Federation of Eugenic Organizations," *Eugenical News* 14, 11 (1929): 153–6.

"The Third International Congress of Eugenics," *Science* 73, 1892 (1931): 357–8.

"Voluntary Sterilization Bill," *The Eugenics Review* 27, 2 (1935): 136–46.

Conferences and Congresses

Transactions of the Seventh International Congress of Hygiene and Demography. Ed. by C. E. Shelly, 13 vols. London: Eyre and Spottiswoode, 1892.

Bericht über den VIII. Internationalen Congress gegen den Alkoholismus. Ed. by Rudolf Wlassak. Vienna: F. Deuticke, 1902.

Bericht über den IX. Internationalen Kongress gegen den Alkoholismus (Bremen 14–19.IV.1903). Ed. by Franziskus Hähnel. Jena: Verlag von Gustav Fischer, 1904.

Internationalen Gesellschaft für Rassen-Hygiene. 5. Bericht, vom 19. März 1909 bis 12. Februar 1910. Naumburg: Lippert. 1910.

Fortpflanzung, Vererbung, Rassenhygiene. Ed. by Max von Gruber and Ernst Rüdin. Munich: J. F. Lehmann, 1911.

Problems in Eugenics. vol. I and II. Papers Communicated to the First International Eugenics Congress held at the University of London, July 24th to 30th, 1912. London: The Eugenics Education Society, 1912 and 1913.

XVIIth International Congress of Medicine, London, 1913. London: Henry Frowde, 1914.

Die Erhaltung und Vermehrung der deutschen Volkskraft. Verhandlungen der 8. Konferenz der Zentralstelle für Volkswohlfahrt in Berlin vom 26. bis 28. Oktober 1915. Berlin: Heymann, 1916.

A népegészségi országos nagygyűlés munkálatai. Ed. by Béla Fenyvessy, József Madzsar. Budapest: Eggenberger, 1918.

Der Wiederaufbau der Volkskraft nach dem Kriege. Jena: Verlag von Gustav Fischer, 1918.

Eugénique et sélection. Ed. by E. Apert et al. Paris: Felix Alcan, 1922.

Scientific Papers of the Second International Congress of Eugenics. 2 vols. Baltimore: Williams & Wilkins, 1923.

Proceedings of the World Population Conference. Ed. by Margaret Sanger. London: Edward Arnold, 1927.

Report of the Ninth Conference of the International Federation of Eugenic Organisations. London: I.F.E.O. 1930.

L'Église et l'eugénisme. La famille a la croisée des chemins. Paris: Éditions Mariage et Famille, 1930.

Proceedings of the First International Congress on Mental Hygiene. 2 vols. Ed. by Frankwood E. Williams. New York: The International Committee for Mental Hygiene, 1932.

Problems of Population being the Report of the Proceedings of the Second General Assembly of the International Union for the Scientific Investigation of Population Problems. Ed. by G. H. L. F. Pitt-Rivers. London: George Allen, 1932.

Scientific Papers of the Third International Congress of Eugenics. Baltimore: Williams & Wilkins, 1934.

Genética, eugenesia y pedagogia sexual. Libro de las primeras jornadas eugénicas españolas. 2 vols. Ed. by Enrique Noguera and Luis Huerta. Madrid: Morata, 1934.

Bevölkerungsfragen. Bericht des Internationalen Kongresses für Bevölkerungswissenschaft. Ed. by Hans Harmsen, Franz Lohse. Munich: J. F. Lehmanns, 1936.

Premier Congrès Latin d'Eugénique. Rapport. Paris: Masson, 1937.

Deliberationes Congressus Dermatologorum Internationalis IX, vol. 2. Budapest: Pátria, 1936; vol. 3. Leipzig: Johann Ambrosius Barth, 1937.

Congrès International de la Population. Paris: Hermann, 1938.

XVIIe Congrès International d'Anthropologie et d'Archéologie Préhistorique. Bucharest: Imprimere Socec, 1939.

General References

Glass, David V. *Population Policies and Movements in Europe.* Oxford: Clarendon Press, 1940.

Hall, Gertrude E. *A Bibliography of Eugenics and Related Subjects.* Albany, N.Y.: Capitol, 1913.

Holmes, Samuel J. *A Bibliography of Eugenics.* Berkeley: University of California Press, 1924.

Martiriul evreilor din România, 1940–1944. Documente şi mărturii. Bucharest: Hasefer, 1991.

Memorial-Volume in Honor of the 100th Birthday of J. G. Mendel. Prague: Fr. Borový, 1925.

Nisot, Marie-Thérèse. *La Question Eugenique dans Les Divers Pays,* 2 vols. Brussels: Falk Fils, 1927, 1929.

Serving the Cause of Public Health. Selected Papers of Andrija Štampar. Ed. by M. D. Grmek. Zagreb: Andrija Štampar School of Public Health, 1966.

Engs, Ruth C. *The Eugenics Movement: An Encyclopaedia.* Westport, Conn.: Greenwood Press, 2005.

Erblehre und Rassenhygiene im völkischen Staat. Munich: J. F. Lehmann, 1934.

Eugenics: Then and Now. Ed. by Carl Jay Bajema. Stroudsburg: Dowden, Hutchinson & Ross, 1976.

Fascism. Ed. by Roger Griffin. Oxford: Oxford University Press, 1995.

The Eugenics Movement: An International Perspective. Ed. by Pauline M. H. Mazumdar. 6 vols. New York: Routledge, 2007.

The Life, Letters and Labours of Francis Galton. Ed. by Karl Pearson. 3 vols. Cambridge: Cambridge University Press, 1914. 1924, 1930.

The Nazi Germany Sourcebook. An Anthology of Texts. Ed. by Roderick Stackelberg, Sally A. Winkle. New York: Routledge, 2002.

The Swedish Nation in Word and Picture, together with short summaries of the contri-bution made by Swedes within the fields of Anthropology, Race-Biology, Genetics and Eugenics. Ed. by H. Lundborg and J. Runnström. Stockholm: Hasse W. Tullberg, 1921.

* * *

SECONDARY LITERATURE

Special Issues Devoted to Eugenics

"Eugenics Old and New," *New Formations* 60 (2007).
"Eugenics and Science," *Science in Context* 11, 3/4 (1998).
"Eugenics, Sex and the State," *Studies in History and Philosophy of Biological and Biomedical Sciences* 39, 2 (2008).
"The Public and Private History of Eugenics," *The Public Historian* 29, 3 (2007).

Articles and Chapters in Edited Volumes

Accampo, Elinor A. "The Gendered Nature of Contraception in France: Neo-Malthusianism, 1900–1920," *Journal of Interdisciplinary History* 34, 2 (2003): 235–62.

Alemdaroğlu, Ayça. "Politics of the Body and Eugenic Discourse in Early Republican Turkey," *Body & Society* 11, 3 (2004): 86–101.

——. "Eugenics, Modernity and Nationalism," in David Turner, Kevin Stagg, eds., *Social Histories of Disability and Deformity* (London: Routledge, 2006). pp. 126–41.

Allen, Ann Taylor. "Mothers of the New Generation: Adele Schreiber, Helene Stocker, and the Evolution of a German Idea of Motherhood, 1900–1914," *Signs* 10, 3 (1985): 418–38.

——. "Feminism and Eugenics in Germany and Britain, 1900–1940: A Comparative Perspective," *German Studies Review* 23, 3 (2000): 477–505.

Barett, Deborah and Charles Kurzman, "Globalizing Social Movement Theory: The Case of Eugenics," *Theory and Society* 33, 5 (2004): 487–527.

Bock, Gisela. "Racism and Sexism in Nazi Germany: Motherhood, Compulsory Sterilization, and the State," *Signs: Journal of Women in Culture and Society* 8, 3 (1983): 400–21.

——. "Sterilization and 'Medical' Massacres in National Socialist Germany: Ethic, Politics, and the Law," in Manfred Berg and Geoffrey Cocks, eds.,

Medicine and Modernity: Public Health and Medical Care in Nineteenth-and Twentieth Century Germany. Cambridge: Cambridge University Press, 1997. pp. 149–72.

——. "Nationalsozialistische Sterilisationpolitik," in Klaus-Dietmar Henke, ed., *Tödliche Medizin im Nationalsozialismus. Von der Rassenhygiene zum Massenmord.* Cologne: Böhlau, 2008. pp. 85–99.

Burleigh, Michael. "Eugenic Utopias and the Genetic Present," *Totalitarian Movements and Political Religions* 1, 1 (2000): 56–77.

Cassata, Francesco. "A 'Scientific Basis' for Fascism: The Neo-Organicism of Corrado Gini," *History of Economic Ideas* 16, 3 (2008): 49–64.

Crook, Paul. "American Eugenics and the Nazis: Recent Historiography," *The European Legacy* 7, 3 (2002): 363–81.

——. "The New Eugenics? The Ethics of Bio-Technology," *Australian Journal of Politics and History* 54, 1 (2008): 135–43.

Davis, Chris. R. "Restocking the Ethnic Homeland: Ideological and Strategic Motives behind Hungary's 'Hazatelepítés' Schemes during WWII (and the Unintended Consequences)," *Regio. Minorities, Politics, Society* 1 (2007): 155–74.

Dickinson, Edward Ross. "Biopolitics, Fascism, Democracy: Some Reflections on Our Discourse about 'Modernity'," *Central European History* 37, 1 (2004): 1–48.

Dikötter, Frank. "Race Culture: Recent Perspectives on the History of Eugenics," *American Historical Review* 103, 2 (1998): 467–78.

Drouard, Alain. "Concerning Eugenics in Scandinavia. An Evolution of Recent Research and Publication," *Population* 11 (1999): 261–70.

——. "Eugenics in France and in Scandinavia: Two Case Studies," in Robert A. Peel, *Essays in the History of Eugenics* (London: The Galton Institute, 1998). pp. 173–207.

Ehrenström, Philippe. "Eugenisme et santé publique: la stérilisation légale des malades mentaux dans le canton de Vaud (Suisse)," *History and Philosophy of the Life Sciences,* 15, 2 (1993): 205–27.

Ekberg, Marryn, "The Old Eugenics and the New Genetics Compared," *Social History of Medicine* 20, 3 (2007): 581–93.

Ergin, Murat. "Biometrics and Anthropometrics: The Twins of Turkish Modernity," *Patters of Prejudice* 42, 3 (2008): 281–304.

Falina, Maria. "Between 'Clerical Fascism' and Political Orthodoxy: Orthodox Christianity and Nationalism in Interwar Serbia," in Matthew Feldman, Marius Turda, eds., *Clerical Fascism in Interwar Europe.* London: Routledge, 2008. pp. 35–46.

Farrall, Lyndsay A. "The History of Eugenics: A Bibliographical Review," *Annals of Science* 36 (1979): 111–23.

Field, Geoffrey. "Nordic Racism," *Journal of the History of Ideas* 38, 3 (1977): 523–40.

Freeden, Michael. "Eugenics and Progressive Thought: A Study in Ideological Affinity," *The Historical Journal* 22, 3 (1939): 645–71.

Gejman, Pablo V., A. Weilbaecher, "History of the Eugenic Movement," *Israel Journal of Psychiatry and Related Sciences* 39, 4 (2003): 217–31.

Gentile, Emilio. "The Myth of National Regeneration in Italy. From Modernist Avant-Garde to Fascism," in Matthew Affron, Mark Antliff, eds., *Fascist Visions*. Princeton: Princeton University Press, 1997. pp. 25–45.

Georgescu, Tudor. "The Eugenic Fortress: Alfred Csallner and the Saxon Eugenic Discourse in Interwar Romania," in Christian Promitzer, Sevasti Trubeta, Marius Turda, eds., *Hygiene, Health and Eugenics in Southeastern Europe to 1945*. Budapest: Central European University Press, 2010.

Gerodetti, Natalia. "From Science to Social Technology: Eugenics and Politics in Twentieth-Century Switzerland," *Social Politics* 13, 1 (2006): 59–88.

Gillette, Aaron, "The Origins of the 'Manifesto of Racial Scientists'," *Journal of Modern Italian Studies* 6, 3 (2001): 305–23.

Graham, Loren R. "Science and Values: The Eugenics Movement in Germany and Russia in the 1920s," *The American Historical Review* 82, 5 (1977): 1133–64.

Griffin, Roger. "Modernity, Modernism, and Fascism. A 'Mazeway Resynthesis'," *Modernism/Modernity* 15, 1 (2008): 9–24.

Güvercin, C. H. and Berna Arda, "Eugenics Concept: From Plato to Present," *Journal of Human Reproduction & Genetic Ethics* 14, 2 (2008): 20–6.

Hall, Leslie A. "Malthusian Mutations: The Challenging Politics and Moral Meanings of Birth Control in Britain," in Brian Dolan, ed. *Malthus, Medicine, and Morality: 'Malthusianism' after 1878*. Amsterdam: Rodopi, 2000. pp. 141–63.

Hansen, Nancy E. and Heidi L. Janz, Dick J. Sobsey, "21st Century Eugenics?" *The Lancet* 372, supplement 1 (2008): 104–7.

Hansen, Randall and Desmond King, "Eugenic Ideas, Political Interests, and Policy Variance. Immigration and Sterilization Policy in Britain and the US," *World Politics* 53, 2 (2001): 237–63.

Hauner, Milan L. "A German Racial Revolution?" *Journal of Contemporary History* 19, 4 (1984): 669–87.

Hietala, Marjatta. "From Race Hygiene to Sterilization: The Eugenics Movement in Finland," in Gunnar Broberg, and Nils Roll-Hansen. eds. *Eugenics and the Welfare State: Sterilization Policy in Denmark, Sweden, Norway, and Finland*. East Lansing: Michigan State University Press, 2005. pp. 195–258.

Jones, Greta. "Eugenics and Social Policy between the Wars," *The Historical Journal* 25, 3 (1982): 717–28.

Jones, Greta. "Eugenics in Ireland: The Belfast Eugenics Society, 1911–1915," *Irish Historical Studies* 28, 109 (1992): 81–95.

Kalling, Ken. "The Self-Perception of a Small Nation: The Reception of Eugenics in Interwar Estonia," in Marius Turda, Paul J. Weindling, eds., *Blood and Homeland: Eugenics and Racial Nationalism in Central and Southeast Europe, 1900–1940*. Budapest: Central European University, 2007. pp. 253–62.

King, Desmond and Randall Hansen, "Experts at Work: State Autonomy, Social Learning and Eugenic Sterilization in 1930s Britain," *British Journal of Political Science* 29, 1 (1999): 77–107.

Koch, Lene. "The Meaning of Eugenics: Reflections on the Government of Genetic Knowledge in the Past and Present," *Science in Context* 17, 3 (2004): 315–31.

——. "Past Futures: On the Conceptual History of Eugenics – a Social Technology of the Past," *Technology Analysis & Strategic Management* 18, 3/4 (2006): 329–44.

——. "Eugenic Sterilisation in Scandinavia," *The European Legacy* 11, 3 (2006): 299–309.

Koven, Seth. "Remembering and Dismemberment: Crippled Children, Wounded Soldiers, and the Great War in Great Britain," *The American Historical Review* 99, 4 (1994): 1167–202.

——. "Eugenic Sterilisation in Scandinavia," *The European Legacy* 11, 3 (2006): 299–309.

Kuechenhoff, Bernhard. "The Psychiatrist Auguste Forel and His Attitude to Eugenics," *History of Psychiatry* 19, 2 (2008): 215–23.

Lepicard, Etienne. "Eugenics and Roman Catholicism: An Encyclical Letter in Context: *Casti Connubii*, December 31, 1930," *Science in Context* 11, 3/4 (1998): 527–44.

Löscher, Monika. "Eugenics and Catholicism in Interwar Austria," in Marius Turda, Paul J. Weindling, eds., *Blood and Homeland: Eugenics and Racial Nationalism in Central and Southeast Europe, 1900–1940*. Budapest: Central European University Press, 2007. pp. 299–313.

Macnicol, John. "Eugenics, Medicine and Mental Deficiency: An Introduction," *Oxford Review of Education* 9, 3 (1983): 177–80.

——. "Eugenics and the Campaign for Voluntary Sterilization in Britain between the Wars," *Social History of Medicine* 2, 2 (1989): 147–69.

——. "The Voluntary Sterilization Campaign in Britain, 1918–39," *The Journal of the History of Sexuality* 2, 3 (1992): 422–38.

Mazumdar, Pauline M. H. "Blood and Soil: The Serology of the Aryan Racial State," *Bulletin of the History of Medicine* 64, 2 (1990): 187–219.

——. "Two Models for Human Genetics: Blood Grouping and Psychiatry in Germany between the World Wars," *Bulletin of the History of Medicine* 70, 4 (1996): 609–57.

McLaren, Angus. "Reproduction and Revolution: Paul Robin and Neo-Malthusianism in France," in Brian Dolan, ed., *Malthus, Medicine, and Morality: "Malthusianism" after 1878*. Amsterdam: Rodopi, 2000, pp. 165–88.

McMahon, Richard. "On the Margins of International Science and National Discourse: National Identity Narratives in Romanian Race Anthropology," *European Review of History* 16, 1 (2009): 101–23.

Mircheva, Gergana. "Marital Health and Eugenics in Bulgaria, 1878–1940," in Christian Promitzer, Sevasti Trubeta, Marius Turda, eds., *Hygiene, Health and Eugenics in Southeastern Europe to 1945*. Budapest: Central European University Press, 2010.

Morrissey, Christof. "Ethnic Politics and Scholarly Legitimation: The German Institut für Heimatforschung in Slovakia, 1941–1944," in Ingo Haar, Michael Fahlbusch, eds., *German Scholars and Ethnic Cleansing, 1919–1945*. New York: Berghahn Books, 2005. pp. 100–9.

Neumann, Boaz. "The Phenomenology of the German People's Body (Volkskörper) and the Extermination of the Jewish Body," *New German Critique* 36, 1 (2009): 149–81.

Nye, Robert. "The Rise and Fall of the Eugenics Empire: Recent Perspectives on the Impact of Biomedical Thought in Modern Society," *The Historical Journal* 36, 3 (1993): 687–700.

Offen, Karen. "Depopulation, Nationalism, and Feminism in Fin-de-Siècle France," *American Historical Review* 89, 3 (1984): 648–76.

Paul, Harry W. "Religion and Darwinism. Varieties of Catholic Reaction," in Thomas F. Glick, ed., *The Comparative Reception of Darwinism*. Chicago: Chicago University Press, 1988. pp. 403–36. [first edition 1974]

Payne, Stanley G. "On the Heuristic Value of the Concept of Political Religion and its Application," *Totalitarian Movements and Political Religions* 6, 2 (2005): 163–74.

Pickens, Donald K. "The Sterilization Movement: The Search for Purity in Mind and State," *Phylon* 28, 1 (1967): 78–94.

Planert, Ute. "Der dreifache Körper des Volkes: Sexualität, Biopolitik und die Wissenschaften vom Leben," *Geschichte und Gesellschaft* 26, 4 (2000): 539–76.

Polsky, Allyson D. "Blood, Race, and National Identity: Scientific and Popular Discourses," *Journal of Medical Humanities* 23, 3/4 (2002): 171–86.

Promitzer, Christian. "Taking Care of the National Body: Eugenic Visions in Interwar Bulgaria, 1905–1940," in Marius Turda, Paul J. Weindling, eds., *Blood and Homeland: Eugenics and Racial Nationalism in Central and Southeast Europe, 1900–1940*. Budapest: Central European University, 2007. pp. 223–52.

Ramsden, Edmund, "Eugenics from the New Deal to the Great Society: Genetics, Demography and Population Quality," *Studies in History and Philosophy of Biological and Biomedical Sciences* 39, 4 (2008): 391–406.

Raz, Aviad E. "Eugenic Utopias/Dystopias, Reprogenetics, and Community Genetics," *Sociology of Health and Illness* 31, 4 (2009): 602–16.

Repp, Kevin. "'More Corporeal, More Concrete': Liberal Humanism, Eugenics, and German Progressives at the Last Fin de Siècle," *The Journal of Modern History* 72, 3 (2000): 683–730.

Roelcke, Volker. "Zeitgeist und Erbgesundheitsgesetzgebung im Europe der 1930er Jahre. Eugenik, Genetik und Politik im historischen Kontext," *Der Nervenarzt* 73, 11 (2002): 1019–30.

Sandall, Roger. "Sir Francis Galton and the Roots of Eugenics," *Society* 45, 2 (2008): 170–6.

Schneider, William H. "Chance and Social Setting in the Application of the Discovery of Blood Groups," *Bulletin of the History of Medicine* 57, 4 (1983): 545–62.

Schwartz, Michael. "Biopolitik in der Moderne. Aspekte eines vielschichtigen 'Machtdispositivs'," *Internationale Wissenschaftliche Korrespondenz zur Geschichte der deutschen Arbeiterbewegung* 31 (1995): 332–47.

Searle, G. R "Eugenics and Politics in Britain in the 1930s," *Annals of Science* 36 (1979): 159–69.

Šimůnek, Michal, "Eugenics, Social Genetics and Racial Hygiene: Plans for the Scientific Regulation of Human Heredity in the Czech Lands, 1900–1925," in Marius Turda, Paul J. Weindling, eds., *Blood and Homeland: Eugenics and Racial Nationalism in Central and Southeast Europe, 1900–1940*. Budapest: Central European University, 2007. pp. 145–66.

Somit, Albert. "Biopolitics," *British Journal of Political Science* 2, 2 (1972): 209–38.

Sonn, Richard. "'Your Body is Yours': Anarchism, Birth Control, and Eugenics in Interwar France," *Journal of the History of Sexuality* 14, 4 (2005): 415–32.

Spektorowksi, Alberto and Elisabet Mizrachi, "Eugenics and the Welfare in Sweden: The Politics of Social Margins and the Idea of a Productive Society," *Journal of Contemporary History* 39, 3 (2004): 333–52.

Spiegel, Gabrielle M. "The Task of the Historian," *American Historical Review* 114, 1 (2009): 1–15.

Stepan, Nancy L. "'Nature's Pruning Hook': War, Race, and Evolution, 1914–18," in J. M. W. Bean, ed., *The Political Culture of Modern Britain: Studies in Memory of Stephen Koss*. London: Hamish Hamilton, 1987. pp. 129–48.

Stone, Dan. "Race in British Eugenics," *European History Quarterly* 31, 3 (2001): 397–425.

Szelényi, Balász A. "From Minority to Übermensch: The Social Roots of Ethnic Conflict in the German Diaspora of Hungary, Romania and Slovakia," *Past and Present* 196, 1 (2007): 215–51.

Trubeta, Sevasti. "Eugenic Birth Control and Prenuptial Health Certificates in Interwar Greece," in Christian Promitzer, Sevasti Trubeta, Marius Turda, eds., *Hygiene, Health and Eugenics in Southeastern Europe to 1945*. Budapest: Central European University Press, 2010.

Turda, Marius. "The Nation as Object: Race, Blood and Biopolitics in Interwar Romania," *Slavic Review* 66, 3 (2007): 413–41.

———. "The First Debates on Eugenics in Hungary, 1910–1918," in Marius Turda, Paul J. Weindling, eds., *Blood and Homeland: Eugenics and Racial Nationalism in Central and Southeast Europe, 1900–1940*. Budapest: Central European University Press, 2007. pp. 185–221.

———. "From Craniology to Serology: Racial Anthropology in Interwar Hungary and Romania," *Journal of the History of Behavioral Sciences* 43, 3 (2007), 361–77.

———. "Recent Scholarship on Race and Eugenics," *The Historical Journal* 51, 4 (2008): 1115–24.

———. "'To End the Degeneration of a Nation': Debates on Eugenic Sterilization in Interwar Romania," *Medical History* 53, 1 (2009): 77–104.

———. "The Biology of War: Eugenics in Hungary, 1914–1918," *Austrian History Yearbook* 40, 1 (2009): 238–64.

———. "Race, Science and Eugenics in the Twentieth Century." In Alison Bashford, Phillipa Levine. eds. *The Oxford Handbook of the History of Eugenics*. New York: Oxford University Press, 2010, pp. 98–127.

Waller, John C. "Ideas of Heredity, Reproduction and Eugenics in Britain, 1800–1875," *Studies in History and Philosophy of Biological and Biomedical Sciences* 32, 3 (2001): 457–89.

Weindling, Paul. "Fascism and Population in Comparative European Perspective," *Population and Development Review* 14 (1988): 102–21.

———. "International Eugenics: Swedish Sterilization in Context," *Scandinavian Journal of History* 24, 2 (1999): 179–97.

———. "A City Regenerated: Eugenics, Race and Welfare in Interwar Vienna," in Deborah Holmes and Lisa Silverman, eds., *Interwar Vienna: Culture between Tradition and Modernity*. New York: Camden House, 2009. pp. 81–113.

Weingart, Peter. "German Eugenics between Science and Politics," *Osiris* 5 (1989): 260–82.

———. "Eugenics – Medical or Social Science?" *Science in Context* 8, 1 (1995): 197–207.

———. "Science and Political Culture: Eugenics in Comparative Perspective," *Scandinavian Journal of History* 24, 2 (1999): 163–77.

Weiss, Sheila Faith. "The Race Hygiene Movement in Germany," *Osiris* 3 (1987): 193–236.

Wilson, Philip K. "Harry Laughlin's Eugenic Crusade to Control the 'Socially Inadequate' in Progressive Era America," *Patterns of Prejudice* 36, 1 (2002): 49–67.

Williams, John Alexander. "Ecstasies of the Yong: Sexuality, the Youth Movement, and Moral Panic in Germany on the Eve of the First World War," *Central European History* 34, 2 (2001): 163–89.

Yeomans, Rory. "Of 'Yugoslav Barbarians' and Croatian Gentlemen Scholars: Nationalist Ideology and Racial Anthropology in Interwar

Yugoslavia," in Marius Turda, Paul J. Weindling, eds., *Blood and Homeland: Eugenics and Racial Nationalism in Central and Southeast Europe, 1900–1940.* Budapest: Central European University Press, 2007. pp. 83–122.

General Works

Adams, Mark B. ed. *The Wellborn Science: Eugenics in Germany, France, Brazil, and Russia.* New York: Oxford University Press, 1990.

Aly, Götz. *Hitler's Beneficiaries: Plunder, Racial War, and the Nazi Welfare State.* London: Macmillan, 2007.

Aly, Götz and Peter Chroust, Christian Pross, *Cleansing the Fatherland. Nazi Medicine and Racial Hygiene.* Baltimore: The Johns Hopkins University Press, 1994.

Arata, Stephen. *Fictions of Loss in the Victorian Fin-de-Siècle.* Cambridge: Cambridge University Press, 1996.

Baader, Gerhard and Veronika Hofer, Thomas Mayer, eds, *Eugenik in Österreich: biopolitischen strukturen von 1900–1945.* Vienna: Czernin Verlag, 2007.

Bashford, Alison. *Imperial Hygiene: A Critical History of Colonialism, Nationalism and Public Health.* Basingstoke: Palgrave Macmillan, 2004.

Bashford, Alison and Phillipa Levine. eds. *The Oxford Handbook of the History of Eugenics.* New York: Oxford University Press, 2010.

Bauman, Zygmunt. *Modernity and the Holocaust.* Cambridge: Polity Press, 1989.

Berman, Marshall. *All that Is Solid Melts into Air: The Experience of Modernity.* New York: Penguin, 1988 [first edition 1982].

Bernardini, Jean-Marc. *Le darwinisme social en France (1859–1918). Fascination et rejet d'une idéologie.* Paris: CNRS Éditions, 1997.

Biddiss, Michael. *The Age of the Masses. Ideas and Society in Europe since 1870.* Harmondsworth: Penguin Books, 1977.

Bock, Gisela. *Zwangssterilisation und Nationalsozialismus: Studien zur Rassenpolitik und Frauenpolitik.* Opladen: Westdeutscher Verlag, 1986.

Bowler, Peter J. *The Mendelian Revolution: The Emergence of Hereditarian Concepts in Modern Science and Society.* London: The Athlone Press, 1989.

Bridenthal, Renate and Atina Grossmann, and Marion Kaplan. eds.. *When Biology became Destiny: Women in Weimar and Nazi Germany.* New York: Monthly Review Press, 1984.

Broberg, Gunnar and Nils Roll-Hansen. eds. *Eugenics and the Welfare State: Sterilization Policy in Denmark, Sweden, Norway, and Finland.* East Lansing: Michigan State University Press, 2005 [first edition 1996].

Bucur, Maria. *Eugenics and Modernization in Interwar Romania.* Pittsburgh: Pittsburgh University Press, 2002.

Burleigh, Michael. *Germany turns Eastwards: A Study of Ostforschung in the Third Reich*. Cambridge: Cambridge University Press, 1988.

Burleigh, Michael and Wolfgang Wippermann, *The Racial State: Germany 1933–1945*. Cambridge: Cambridge University Press, 1991.

Byer, Doris. *Rassenhygiene und Wohlfahrtspflege. Zur Entstehung eines sozialdemokratischen Machtdispositivs in Österreich bis 1934*. Frankfurt: Campus Verlag, 1988.

Carol, Anne. *Histoire de l'eugénisme en France: Les médecins et la procréation, XIXe-XXe siècle*. Paris: Seuil, 1995.

Cassata, Francesco. *Molti, Sani e Forti. L'eugenetica in Italia*. Turin: Bollati Boringhieri, 2006.

——. *"La Difesa della razza". Politica, ideologia e immagine del razzismo fascista*. Torino: Giulio Einaudi, 2008.

Chamberlain, Edward J. and Sander L. Gilman, *Degeneration: The Dark Side of Progress*. New York: Columbia University Press, 1985.

Childers, Kristen Stromberg. *Fathers, Families, and the State, 1914–1945*. Ithaca: Cornell University Press, 2003.

Cleminson, Richard. *Anarchism, Science and Sex: Eugenics in Eastern Spain, 1900–1937*. Bern: Peter Lang, 2000.

Connelly, Matthew. *Fatal Misconception: The Struggle to Control World Population*. Cambridge, Mass.: Harvard University Press, 2008.

Crook, Paul. *Darwinism, War and History. The Debate over the Biology of War from the "Origin of Species" to the First World War*. Cambridge: Cambridge University Press, 1994.

Deák, István. *Beyond Nationalism: A Social and Political History of the Habsburg Officer Corps, 1848–1918*. New York: Oxford University Press, 1990.

Dickinson, Edward Ross. *The Politics of German Child Welfare from the Empire to the Federal Republic*. Cambridge, Mass.: Harvard University Press, 1996.

Dolan, Brian. ed. *Malthus, Medicine, and Morality "Malthusianism" after 1798*. Amsterdam: Rodopi, 2000.

Dowbiggin, Ian. *The Sterilization Movement and Global Fertility in the Twentieth Century*. New York: Oxford University Press, 2008.

Drouard, Alain. *Une inconnue des sciences sociales. La Foundation Alexis Carrel, 1941–1945*. Paris: Éditions de la Maison des Sciences de l'Homme, 1992.

——. *L'eugénisme en questions: L'exemple de l'eugénisme français*. Paris: Ellipses, 1999.

Dwork, Deborah. *War is Good for Babies. A History of the Infant and Child Welfare Movement in England, 1898–1918*. London: Tavistock Publications, 1987.

Ekstein, Modris. *Rites of Spring: The Great War and the Birth of the Modern Age*. London: Macmillan, 2000 [first edition 1989].

Esposito, Roberto. *Bios: Biopolitics and Philosophy*. Minneapolis: University of Minnesota Press, 2008.

SELECT BIBLIOGRAPHY

Evans, Richard J. *The Third Reich in Power, 1933–1939*. London: Allen Lane, 2005.

Feldmann, Matthew and Marius Turda, eds., *Clerical Fascism in Interwar Europe*. London: Routledge, 2008.

Foucault, Michel. *Discipline and Punish: The Birth of the Prison*. New York: Vintage, 1977.

——. *The History of Sexuality: An Introduction*. New York: Vintage Books, 1980.

——. *"Society Must be Defended". Lectures at the Collège de France, 1975–1976*. New York: Picador, 1997.

Fritzsche, Peter. *Germans into Nazis*. Cambridge, Mass.: Harvard University Press, 1998.

——. *Stranded in the Present: Modern Time and the Melancholy of History*. Cambridge, Mass.: Harvard University Press, 2004.

Gawin, Magdalena. *Rasa i nowoczesność: historia polskiego ruchu eugenicznego, 1880–1952*. Warsaw: Wydawnicwo Neriton, 2003.

Gay, Peter. *Modernism: The Lure of Modernism from Baudelaire to Becket and Beyond*. London: William Heinemann, 2007.

Gentile, Emilio. *The Struggle for Modernity: Nationalism, Futurism, and Fascism*. Westport, Conn.: Praeger, 2003.

Gillette, Aaron. *Racial Theories in Fascist Italy*. New York: Routledge, 2001.

——. *Eugenics and the Nature-Nurture Debated in the Twentieth Century*. New York: Palgrave Macmillan, 2007.

Goffman, Erving. *Stigma. Notes on the Management of Spoiled Identity*. New York: Simon & Schuster, 1986 [first edition 1963].

Griffin, Roger. *Modernism and Fascism: The Sense of a Beginning under Mussolini and Hitler*. Basingstoke: Palgrave Macmillan, 2007.

Griffin Roger and Robert Mallett, John Tortorice, *The Sacred in Twentieth-Century Politics*. Basingstoke: Palgrave Macmillan, 2008.

Grossmann, Atina. *Reforming Sex: The German Movement for Birth Control and Abortion Reform, 1920–1950*. Oxford: Oxford University Press, 1995.

Hanebrink, Paul A. *In Defense of Christian Hungary. Religion, Nationalism, and Antisemitism, 1890–1944*. Ithaca: Cornell University Press, 2006.

Heller, Geneviève and Gilles Jeanmonod, Jacques Gasser, *Rejetées, rebelles, mal adaptées: débats sur l'eugénisme : pratiques de la stérilisation non volontaire en Suisse romande au XXe siècle*. Geneva: Georg, 2002.

Hong, Young-Sun. *Welfare, Modernity, and the Weimar State, 1919–1933*. Princeton: Princeton University Press, 1998.

Horn, David G. *Social Bodies: Science, Reproduction, and Italian Modernity*. Princeton: Princeton University Press, 1994.

——. *The Criminal Body: Lombroso and the Anatomy of Deviance*. London: Routledge, 2003.

Hutton, Christopher M. *Race and the Third Reich*. Cambridge: Polity Press, 2005.

Ipsen, Allan. *Dictating Demography: The Problem of Population in Fascist Italy*. Cambridge: Cambridge University Press, 2002.

Jones, Greta. *Social Hygiene in Twentieth Century Britain*. London: Croom Helm, 1986.

Kallis, Aristotle. *Genocide and Fascism. The Eliminationist Drive in Fascist Europe*. London: Routledge. 2009.

Kevles, Daniel. *In the Name of Eugenics: Genetics and the Uses of Human Heredity*. Cambridge, Mass.: Harvard University Press, 1985.

Koch, Lene. *Racehygiejne i Danmark, 1920–56*. Copenhagen: Gyldendal, 1996.

Komjáthy, Anthony and Rebecca Stockwell, *German Minorities and the Third Reich: Ethnic Germans of East Central Europe between the Wars*. New York: Holms and Meier, 1980.

Koonz, Claudia. *Mothers in the Fatherland: Women, the Family and Nazi Politics*. New York: St Martin's Press – now Palgrave Macmillan, 1987.

Koselleck, Reinhardt. *Futures Past: On the Semantics of Historical Time*. Cambridge, Mass.: MIT Press, 1985.

——. *The Practice of Conceptual History: Timing History, Spacing Concepts*. Stanford: Stanford University Press, 2002.

Kühl, Stefan. *The Nazi Connection: Eugenics, American Racism and German National Socialism*. New York: Oxford University Press, 1994.

——. *Die Internationalen der Rassisten: Aufstieg und Niedergang der Internationalen Bewegung für Eugenik und Rassenhygiene im 20. Jahrhundert*. Frankfurt: Campus Verlag, 1997.

Largent, Mark A. *Breeding Contempt: The History of Coerced Sterilization in the United States*. New Brunswick, NJ.: Rutgers University Press, 2008.

Leotis, Artemis. *Topographies of Hellenism: Mapping the Homeland*. Ithaca: Cornell University Press, 1995.

Link, Gunther. *Eugenische Zwangssterilisationen und Schwangerschaftsarbbrüche im Nationalsozialismus*. Frankfurt am Main: Peter Lang, 1999.

Löscher, Monika. *"... der gesunden Vernuft nicht zuwider ..."? Katholische Eugenik in Österreich vor 1938*. Innsbruck: Studienverlag, 2009.

Lumans, Valdis O. *Himmler's Auxiliaries: The Volksdeutsche Mittelstelle and the German National Minorities of Europe, 1933–1945*. Chapel Hill: University of North Carolina Press, 1993.

MacMaster, Neil. *Racism in Europe, 1870–2000*. Basingstoke: Palgrave – now Palgrave Macmillan, 2001.

Mantovani, Claudia. *Rigenerare la società: l'eugenetica in Italia dalle origini ottocentesche agli anni Trenta*. Catanzaro: Rubbettino, 2004.

Mattila, Markku. *Kansamme parhaaksi. Rotuhygienia Suomessa vuoden 1935 sterilointilakiin asti*. Helsinki: Suomen Historiallinen Seura, 1999.

Morton, Stephen and Stephen Bygrave. eds. *Foucault in an Age of Terror: Essays on Biopolitics and the Defence of Society*. Basingstoke: Palgrave Macmillan, 2008.

Mosse, George L. *Fallen Soldiers: Reshaping the Memory of the World Wars*. Oxford: Oxford University Press, 1990.

——. *The Fascist Revolution. Toward a General Theory of Fascsim*. New York: Howard Fertig, 1999.

Mottier, Véronique and Laura von Mandach. eds. *Eugenik und Disziplinierung in der Schweiz: Integration und Ausschluss in Psychiatrie, Medizin und Fürsorge*. Zürich: Seismo, 2007.

Mouton, Michelle. *From Nurturing the Nation to Purifying the Volk: Weimar and Nazi Family Policy, 1918–1945*. Cambridge: Cambridge University Press, 2007.

Noordman, Jan. *Om de kwaliteit van het nageslacht. Eugenetika in Nederland, 1909–1950*. Nijmegen: SUN, 1989.

Olson, Richard G. *Science and Scientism in Nineteenth-Century Europe*. Champaign, IL.: University of Illinois Press, 2008.

Petrakis, Marina. *The Metaxas Myth: Dictatorship and Propaganda in Greece*. London: I. B. Tauris, 2006.

Pick, Daniel. *Faces of Degeneration: A European Disorder, c.1848–1918*. Cambridge: Cambridge University Press, 1989.

Paul, Diane B. *Controlling Human Heredity: 1865 to the Present*. Atlantic Highlands, NJ: Humanities Press, 1995.

——. *The Politics of Heredity: Essays on Eugenics, Biomedicine, and the Nature-Nurture Debate*. New York: State University of New York Press, 1998.

Peel, Robert A. ed. *Essays in the History of Eugenics*. London: The Galton Institute, 1998.

Porter, Roy. *Bodies Politic: Disease, Death and Doctors in Britain, 1650–1900*. Ithaca: Cornell University Press, 2001.

Presner, Todd. *Muscular Judaism: The Jewish Body and the Politics of Regeneration*. New York: Routledge, 2007.

Promitzer, Christian and Sevasti Trubeta, Marius Turda, eds., *Hygiene, Health and Eugenics in Southeastern Europe to 1945*. Budapest: Central European University Press, forthcoming 2010.

Prum, Michel. ed. *Exclure au nom de la race : Étas-Unis, Irlande, Grande-Bretagne*. Paris: Syllepse, 2000.

Quine, Maria Sophia. *Italy's Social Revolution. Charity and Welfare from Liberalism to Fascism*. Basingstoke: Palgrave Macmillan, 2002.

Richardson, Angelique. *Love and Eugenics in the Late Nineteenth Century. Rational Reproduction and the New Woman*. Oxford: Oxford University Press, 2003.

Richter, Ingrid. *Katholizismus und Eugenik in der Weimarer Republic und im Dritten Reich: Zwischen Sittlichkeitsreform und Rassenhygiene*. Paderborn: Ferdinand Schöningh, 2001.

Richter, Melvin. *The History of Political and Social Concepts: A Critical Introduction.* Oxford: Oxford University Press, 1995.

Roberts, David. *The Totalitarian Experiment in Twentieth Century Europe. Understanding the Poverty of Great Politics.* New York: Routledge, 2005.

Runcis, Maija. *Steriliseringar i folkhemmet.* Stockholm: Ordfront, 1998.

Schleiermacher, Sabine. *Sozialethik im Spannungsfeld von Sozial- und Rassenhygiene: Der Mediziner Hans Harmsen im Centralausschuss fur die Innere Mission.* Husum: Matthiesen, 1998.

Schmuhl, Hans-Walter. *Rassenhygiene, Nationalsozialismus, Euthanasie. Von der Verhütung zur Vernichtung "lebensunwerten Lebens", 1890–1945.* Göttingen: Vandenhoeck & Ruprecht, 1987.

——. *The Kaiser Wilhelm Institute for Anthropology, Human Heredity, and Eugenics, 1927–1945.* Dordrecht: Springer, 2008.

Schneider, William H. *Quality and Quantity: The Quest for Biological Regeneration in Twentieth-Century France.* Cambridge: Cambridge University Press, 1990.

Schorske, Carl E. *Thinking with History: Explorations in the Passage to Modernism.* Princeton: Princeton University Press, 1998.

Schweizer, Magdalena. *Die psychiatrische Eugenik in Deutschland und in der Schweiz zur zeit des Nationalsozialismus.* Bern: Peter Lang, 2002.

Searle, G. R. *Eugenics and Politics in Britain, 1900–1914.* Leyden: Noordhoff International Publishing, 1976.

Solonari, Vladimir. *Purifying the Nation: Population Exchange and Ethnic Cleansing in Nazi-Allied Romania.* Baltimore: The Johns Hopkins University Press, 2009.

Soloway, Richard. *Birth Control and the Population Question in England, 1877–1930.* Chapel Hill: The University of North Carolina Press, 1982.

——. *Demography and Degeneration. Eugenics and the Declining Birthrate in Twentieth-Century Britain.* Chapel Hill: The University of North Carolina Press, 1990.

Spring, Claudia Andrea. *Zwischen Krieg und Euthanasie. Zwangssterilisation in Wien, 1940–1945.* Vienna: Böhlau Verlag, 2009.

Stepan, Nancy Leys. *"The Hour of Eugenics": Race, Gender, and Nation in Latin America.* Ithaca: Cornell University Press, 1991.

Stone, Dan. *Breeding Superman: Nietzsche, Race and Eugenics in Edwardian and Interwar Britain.* Liverpool: Liverpool University Press, 2002.

——. ed. *The Historiography of the Holocaust.* Basingstoke: Palgrave Macmillan, 2005.

Thomson, Mathew. *The Problem of Mental Deficiency. Eugenics, Demography, and Social Policy in Britain, c. 1870–1959.* Oxford: Oxford University Press, 1998.

Todorov, Tzvetan. *Imperfect Garden: The Legacy of Humanism.* Princeton: Princeton University Press, 2002.

Turda, Marius. *Eugenism si antropologie rasiala în România, 1874–1944.* Bucharest: Cuvântul, 2008.

——. *A Healthy Nation: Eugenics, Race and Biopolitics in Hungary, 1904–1944.* Budapest: Central European Univesity Press, forthcoming 2011.

Turda, Marius and Paul J. Weindling. eds. *Blood and Homeland: Eugenics and Racial Nationalism in Central and Southeast Europe, 1900–1940.* Budapest Central European University, 2007.

Turda, Marius and Diana Mishkova, eds., *Anti-Modernism: Radical Revisions of Collective Identity.* Budapest: Central European University Press, forthcoming 2011.

Weikart, Richard. *Hitler's Ethic: The Nazi Pursuit of Evolutionary Progress.* New York: Palgrave Macmillan, 2009.

Weindling, Paul. *Health, Race and German Politics between National Unification and Nazism, 1870–1945.* Cambridge: Cambridge University Press, 1989.

Weiner, Amir. ed. *Landscaping the Human Garden. Twentieth-Century Population Management in a Comparative Framework.* Stanford: Stanford University Press, 2003.

Weiss, Sheila Faith. *Racial Hygiene and National Efficiency: The Eugenics of Wilhelm Schallmayer.* Berkeley: University of California Press, 1987.

Wolf, Maria A. *Eugenische Vernuft. Eingriffe in die reproduktive Kultur durch die Medizin, 1900–2000.* Vienna: Böhlau Verlag, 2008.

Zahra, Tara. *Kidnapped Souls: National Indifference and the Battle for Children in the Bohemian Lands, 1900–1948.* Ithaca: Cornell University Press, 2008.

Zarifopol-Johnston, Ilinca. *Searching for Cioran.* Bloomington: Indiana University Press, 2009.

INDEX

abortion, 58, 89, 112
acquired characteristics theory, 17
activism, eugenic, 73–5, 86–7
aesthetics, 2, 78–9
Albania, 102–3
alcoholism, 26, 94
Alemdaroğlu, Ayça, 105
American eugenicists, 33, 38,
    70–1, 80
anarchism, 40, 80
ancestral heredity, law of, 16–17
Anglo-Saxon eugenics, 98–9
Antal, Lajos, 114
anti-Semitism, 45, 93, 110, 115
Antliff, Mark, 120
Antonescu, Ion, 105
Antonescu, Mihai, 116
Apáthy, István, 35–6, 61
artificial selection, 26, 33, 38, 42,
    44, 64, 79, 94
associations and societies, 18, 23,
    33–4, 37–8, 51, 59–61, 73–6,
    80–1, 88, 96–7, 107–8, 114
Austria, 48, 74, 88–9
Austrian eugenicists, 52, 68, 82
Austria-Hungary, 36
    Austro-Hungarian
    Monarchy, 45

Balás, Károly, 76, 110
Banu, Gheorghe, 99
Bársony, János, 58
Bashford, Alison, 5
Bauman, Zygmunt, 116, 120
Baur, Erwin, 69–70
Benedek, László, 84

Benjamin, Walter, 2
Benn, Gottfried, 94
Berman, Marshall, 70
Binding, Karl, 79
biological identity, 2, 6, 8, 18, 66
biological uniqueness theories, 18
biologisation of national
    belonging, 6–8, 67, 77,
    105–6
biologism, 114
biopolitics, 2, 5, 11, 92–3, 122
    and control of ethnic
    minorities, 107–10
    and practical applications of
    eugenics, 93–100
    and regeneration, 100–7
    and the state, 110–17
birth control, 29–30, 58, 89–90
    see also reproduction control
birthrate issues, 27, 57–60
blending inheritance theory, 16
blood groups, 18, 102, 110
Blum, Agnes, 38
Bock, Gisela, 95
body, nation as, 5–6, 64, 66
Boer War, 41
Bossi, Luigi Maria, 58–9
Bottai, Giuseppe, 113
Bowler, Peter, 15, 119
Britain, 28, 60, 72, 84, 96
British eugenicists, 27–8, 33, 36, 38,
    42, 44
Brouillard, René, 88
Bucur, Maria, 82
Bulffi, Luis, 30
Bulgaria, 85, 96

Burleigh, Michael, 15, 120
Buzoianu, Gheorghe, 96

Capitan, Louis, 47–8
Carrel, Alexis, 13
Catholic Church, 85, 88–90
Chambers Theodore, 44
charity, 21–2, 34, 46, 111
Childers, Kristen Stromberg, 59
Christian morality, 84–90
chronic diseases, 29
Cioran, Emil, 93
class differentials, 19, 21, 28–9, 31
Codreanu, Cornelius Zelea, 101,
    104–5
colonies, 29, 32, 39, 54, 81–2
communism, 2, 15
consensus, 3, 13, 49, 63, 72
Crainic, Nichifor, 101–2
criminality, 28, 38, 48, 58, 65–6, 74,
    83, 87, 97, 112
Croatian eugenicists, 69
Csallner, Alfred, 87, 108
Csángós, 109
cultural contexts, 3, 5, 8, 116,
    119–20, 123
cultural identity, 107–8
Czech eugenicists, 31, 33, 36, 65,
    70, 73
Czechoslovakia, 59, 74, 96–7
Czechoslovak Institute of
    National Eugenics, 74

Darányi, Gyula, 98
Darré, Richard Walther, 94–5
Darwin, Charles, 15–16, 19, 46, 125
Darwin, Leonard, 37, 49
Darwinism, 20, 46, 83, 88, 120
Deák, István, 45
decadence, 7, 25, 38, 42–3, 94, 101
definitions of eugenics, 3, 19–22
degenerates, 28, 65–6, 71, 74,
    80, 82

degeneration, 24–31, 35, 73, 82,
    121, 124
degenerative modernity, 4, 6–7,
    25–6, 38, 92, 94, 108
demography, 38, 52, 57
Denmark, 83
depopulation, 54, 62, 87
Dermine, Jean, 88
Devaldès, Manuel, 80
Dickinson, Edward Ross, 112
Dubourg, Monsignor, Bishop of
    Marseilles, 88
dysgenic effects of war, 42, 46–52
dysgenic groups, 11–12, 85, 99, 115

early twentieth century, 5–6, 8, 10,
    12–21, 23, 26–34, 37–8,
    40–1, 121
Eastern Europe, 4, 30–1, 109
    see also individual countries
economic factors, 4, 69, 71, 79, 84
education, 32, 73–4, 77, 89,
    104–5, 107
Eksteins, Modris, 61
Elderton, Ethel, 27
elimination of dysgenics, 11, 33, 42,
    44, 55, 66, 80, 93, 109, 111–13,
    115, 124
Ellis, Havelock, 7, 14, 29
Encyclical on Christian Marriage
    (1930), 89–90
Enlightenment, 89, 120
environmental factors, 16,
    19, 26, 50
Estonian eugenicists, 98, 107
    Estonian Eugenics Society, 74
ethnic minorities, 106–10, 115–16,
    124–5
eugenic institutions, 28–9, 59, 74,
    84, 96, 109, 115
    see also associations and
    societies
eugenic entopia, 100, 106

eugenic practices, 20, 24, 28, 31, 35, 72
  legitimisation of, 67–8, 84
  negative, 10, 38, 51, 70–1, 80–2, 88–9, 99
  positive, 10, 32, 51, 70–1, 88, 116
eugenic stigma, 65–7, 122
  stigmatisation, 66, 69
Eugenics Education Society, 74
Eugenics Record Office, America, 33
euthanasia, 7, 71, 79–83

Făcăoaru, Iordache, 105–6
family policies, 57–60, 76, 113
  see also marriage regulation; reproduction control
fascism, 2, 15, 64, 78, 100–2, 116, 121
fascist aesthetics, 78–9
Federley, Harry, 107–8
feeblemindedness, 28–9, 65, 83, 87, 94, 96–7, 99
feminism, 57–9
fertility issues, 29–30
Filandros, Konstantinos, 85
Finland, 97, 107–8
Finnish eugenicists, 67
First World War, 8, 71, 121
  modernist attitudes to, 40–1
  new era discourse, 62–3
  and political biology, 56–61
  and qualitative vs. quantitative eugenics, 52–6
  and racial damage, 46–52
  as regenerative, 42–6
Fischer, Eugen, 78, 94, 113
Forel, Auguste, 65–6
Foucault Michel, 6, 10, 116
Foustka, Břetislav, 31
France, 29, 59, 88, 98, 115
French eugenicists, 25, 28–30, 33, 38, 40, 47–8, 80, 98, 115
Frick, Wilhelm, 111

Fritzsche, Peter, 94, 120
future generations orientation, 49–50, 59–61, 68

Galton, Francis, 8–9, 16–17, 19–23, 32–3, 37, 39, 44, 49, 67, 76, 84, 90, 116, 126
Gasset, José Ortega, 13
gender issues, 57–9
genealogy, 17, 68, 74, 108
genetic engineering, 12, 123
Gentile, Emilio, 41, 43–4, 64
Georgescu, Tudor, 87
German eugenicists, 26, 29, 36, 38
  First World War, 40, 44, 47–8, 51, 57–8
  interwar years, 69–70, 76–7, 79, 94
German minorities, 108–9
German model, 97–8, 103
German Society for Racial Hygiene, 74
Germany, 29, 56–8, 69–72, 86, 88–95, 105, 111, 113
Gini, Corrado, 38, 49, 98–9, 105
Goffman, Ervin, 66–7
Gökay, Fahrettin Kerim, 98
Grant, Madison, 80, 111
Greater Romania, 11, 101
Greece, 96, 103, 106
Greek eugenicists, 76–7, 85, 104
Griffin, Roger, 2, 6, 25, 120
Gross, Walter, 95
Grotjahn, Alfred, 32, 77
Gruber, Max von, 47
Gruia, Ioan, 110

Harmsen, Hans, 86–7
Haro, Francisco, 96
Harris, G. W., 112
Haškovec, Ladislav, 36, 65
Haţieganu, Iuliu, 78
Hauptmann, Gerhard, 29

Hayek, Friedrich A., 14
healthcare (mothers and infants),
  57–60
health certificates, matrimonial,
  72–3, 95
health courts, 94–5
health reform, 23, 47, 56, 60, 69,
  73, 76, 78
healthy nation vision, 4, 7, 15, 20,
  31, 56
Henlein, Konrad, 108
hereditary diseases, 30, 36, 54,
  65–6, 74, 76, 83
  legislation, 94–5, 97, 111
Herseni, Traian, 114
Hirszfeld, Ludwik, 18
historical research, 108–9
Hitler, Adolf, 94, 105
Hoche, Alfred, 79
Hoffmann, Géza von, 50–1, 60, 70
homogeneity as a goal, 11, 68,
  73–4, 91, 105, 109
Horn, David, 8
Houssay, Frederic, 38
human improvement, 1, 12–13,
  20–1, 41, 45, 77, 81, 83–4, 87,
  123, 125
humanism, 34, 79, 84
Hungarian eugenicists, 34–6, 43,
  50, 58, 60–3, 70, 84, 98, 101–2,
  110, 115
Hungarian Institute for National
  Biology, 115
Hungarian Society for Racial
  Hygiene and Population
  Policy, 74
Hungary, 87, 89, 97, 115
Hygiene Exhibition, Dresden
  (1911), 33, 37

ideals, eugenic, 19–24, 37, 73
identity, 2, 6–8, 18, 66, 69, 92, 102,
  107–8, 122, 125

immigration control, 33
individual/collective relationship,
  4–8, 21–2, 35–6, 73, 76, 92,
  98–9, 113, 120
individualism, 35–6, 38
individuals, relative value of, 12,
  49, 51, 78–80
inheritance, diverse laws of, 16
Innere Mission, 86–7
institutionalisation, 33, 73, 75,
  108–9, 115, 121
institutional support, 56–7, 61
institutions, eugenic, 28–9, 59, 74,
  84, 96, 109, 115
  see also associations and
    societies
intermarriage, 28, 33
International Congress for Racial
  Hygiene, 122
International Congress of
  Medicine, 65
International Congress of Mental
  Hygiene, 84
International Eugenics Congress,
  9–10, 36–7, 62, 122
International Federation of
  Eugenic Organisations, 78,
  99–100
internationalisation, 23, 31–9, 122
International Latin Eugenics
  Congress, 98
interwar years, 8, 11, 64–5, 92–3, 123
  control of ethnic minorities,
    107–10
  nationalisation of eugenics,
    72–9
  post-war reconstruction, 69–72
  practical applications of
    eugenics, 93–100
  religious support/opposition,
    84–91
  state-oriented eugenics, 110–17
  sterilisation, 83–4

interwar years – *continued*
  stigmatisation, 11
Irmak, Sadi, 96, 98
Italian eugenicists, 28–9, 38, 40,
    44, 46, 48–9, 102, 105, 109–10
Italian Eugenics Committee, 33
Italian Society for Genetics and
    Eugenics, 74
Italy, 58–9, 71, 94, 100, 113

Jews, 44–5, 62, 93, 115–16
Jojkić, Mladen, 30
journals, 22, 29–30, 32, 34, 47, 58,
    69–70, 75–6, 83, 95, 106, 112

Kalling, Ken, 107
Kallis, Aristotle, 120
Kaup, Ignaz, 68
Kiss, Géza, 87
Koch, Lene, 3
Kopp, Marie E., 95
Kosellech, Reinhardt, 6
Koumaris, Ioannis, 104

Laanes, Theophil, 98
Lamarck, Jean Baptiste, 15–16
Landra, Guido, 102
Landsteiner, Karl, 18
Lankester, Edwin Ray, 14
Lapouge, Georges Vacher de, 111
La Rochelle, Pierre Drieu, 78
Latin eugenics, 5, 98–9
Lattes, Leone, 102
Latvia, 97
Laughlin, Harry, 69
legislation, 10, 38, 83, 85, 94–9,
    110–11, 115, 121–2
legitimisation, 9, 67–8, 84
Lenz, Fritz, 94
Leontis, Artemis, 106
Lindsay, J. A., 42–3
Lombroso, Cesare, 24, 28
Löscher, Monika, 88–9

Lucidi, Giuseppe, 109–10
Lundborg, Herman, 77–8, 81–2
Lusatian Sorbs, 109

MacMaster, Neil, 57
Macnicol, John, 72
Madrazo, Enrique, 26–7, 35
Madzsar, József, 34, 36
Magyars, 58, 61–3
Maier, Hans, 96
Manifesto of Racial Scientists, 102
Manliu, Ioan, 82, 86
marginalisation, 12, 26, 106–7
Marinetti, Filippo, 40
marriage regulation, 36, 68, 74, 85,
    95–6, 98
Mărtinaş, Ioan, 109
Masaryk, Tomáš Garrigue, 73
matrimonial health certificates,
    72–3, 95
Mayer, Joseph, 89
Méhely, Lajos, 43–4, 101
Mendel, Gregor, 15–16
Mendelian laws of heredity, 16, 18,
    65, 102, 120
mental disability, 67, 83–4, 94
messianism, 93
Metaxas, Ioannis, 103–4
methodologies, 2–3, 5, 118–19, 124
minority group eugenics, 107–9
Mircheva, Gergana, 85–6
Mitchell, Peter Chalmers, 18
Mjøen, Jon Alfred, 32–3, 99–100
modernism, 1–2, 4–6, 25, 105–6,
    118–20
  and national palingenesis, 18,
    109, 123–4
  and new era/improvement
    narratives, 6, 10, 32, 71, 110
  and political revolution, 64–5,
    70, 78, 110, 116
modernist attitudes to war, 8,
    40–1, 51–2, 61

modernist intellectuals, 10, 93–4
modernist version of eugenics, 7, 13
modernity, 120–1
    degenerative, 4, 6–7, 25–6, 38, 92, 94, 108
Moisidis, Moisis, 96
Moldovan, Iuliu, 76–7, 112–13
morality of eugenics, 17–18, 21–4, 29, 31, 34, 57, 61, 79, 82, 112, 116
Morel, Bénédict August, 28
Mosse, George L., 63
motherhood, cult of, 58–9, 108
Muckermann, Hermann, 89–91
Mügge, Maximilian, 22
Mussolini, Benito, 78, 94, 100–2, 104–5
mythologisation, 95, 100–7, 115

National Baby Week, England, 60
national contexts, 4–5, 8, 31–7, 67–8, 118, 123
national efficiency, 41, 82, 121
national eugenics, 10, 32, 36–7, 72–9, 90
National Foundation for the Protection of the Family, Hungary, 98
national identity, 102, 122, 125
national improvement, 22, 32, 111, 122, 125
    technologies, 64–5;
    euthanasia, 7, 71, 79–83; and nationalisation, 72–9; and post-war reconstruction, 69–72; religious support/ opposition, 84–91; sterilisation, 32, 34–5, 38, 80–4, 89, 94–6
nationalism, 1, 3, 9, 11, 15, 39
    First World War, 40, 43, 45, 51, 53–5, 59–60, 62–4

in interwar years, 69, 72, 75, 101–2, 104, 108, 111, 115
national palingenesis, 6, 18, 80, 100–1, 109, 123–4
national protectionism, 10–12
National Socialism, 87, 90, 93, 95, 109, 113
national survival rhetoric, 42, 51, 57, 109
nation as race, 100–1, 104
nation-states, 11–12, 106–8, 121, 125
natural selection, 34, 42, 46
nature *vs.* nurture debate, 15–17, 26
Nazi European Empire, 103
Nazism, 15, 92–5, 104, 116, 120
negative eugenic practices, 10, 38, 51, 70–1, 80–2, 88–9, 99
neo-Malthusianism, 80
new era discourse, 6, 10, 32, 62–3, 71, 110
Nicolai, Georg Friedrich, 51
Niedermeyer, Albert, 89
Nietzsche, Friedrich, 22
nineteenth century, 5–11, 13–17, 19–20, 22, 24–5, 27–30, 37, 46, 68, 83, 107
Nordau, Max, 24–5
normative aims, 66–7, 79
Norway, 97, 103, 115
Norwegian eugenicists, 32–3, 99
Nye, Robert, 3

Olson, Richard, 14
Orthodox Church, 85–7
Ottoman Empire, 30

palingenesis, 6, 18, 25, 43, 45, 80, 86, 100–1, 109, 123–4
    see also regeneration
Pal, Iosif Petru, 109
Panaitescu, Petre P., 106

patriotism, 29, 38, 42–5, 49, 53, 55, 59, 62, 86, 88
Paul, Harry, 88
Pearl, Raymond, 70–1
Pearson, Karl, 16, 26–7, 36–7, 41, 76–7
Péguy, Charles, 29
perfection rhetoric, 1, 11–12, 22, 66, 115, 120
Petit, Eugen, 96
Petrini-Galatzi, Mihail, 30
philanthropy, 21–2
physical degeneration, 27–9
Pick, Daniel, 28
Pinard, Adolphe, 38
Pius XI, 89–90
Ploetz, Alfred, 20, 23, 26, 29, 32–3, 36, 38, 46
Poisson, George, 47
Polakovič, Štefan, 103
Polish Eugenics Society, 74
political biology, 56–61, 68–9, 73
    see also biopolitics
politicisation, 7, 29, 62, 75, 77–8, 86
popularisation, 13–14, 20–1, 33, 37–8, 56, 60, 75, 124
population decrease/increase, 27, 41–2, 48, 50–1, 54–5, 57, 62–3, 112, 121
positive eugenic practices, 10, 32, 51, 70–1, 88, 116
practical applications, 7, 64–5, 77, 93–100
    see also eugenic practices
premarital examination, 115
Presner, Todd, 44
private/public boundaries, 60, 69, 92, 98
professionalisation, 33
programmatic modernism, 2, 22, 27, 38
propaganda, 32, 48, 59–60, 77, 108

prophylactic racial hygiene, 32–3, 76
protectionism, 21, 66, 95
    racial, 11, 29, 51, 68, 70, 76, 82–3, 87, 98, 107–9, 115
    social, 29, 51–3
Protestant Churches, 86–7
public health, 2, 10, 35–6, 65, 68, 70, 95, 107–8, 118
public hygiene, 35–6, 98
puericulture, 9, 32, 38
purification, 15, 27, 38, 95, 116

qualitative vs. quantitative eugenics, 33, 50–6, 114–15
quantitative measurements, 19
Quisling, Vidkun, 103, 115

race-consciousness, 61–2, 76, 81, 92–4
racial damage, wartime, 46–52
racial hygiene, 20, 23, 31–3, 35, 38, 94–5
racial identity, 7, 92
racial improvement, 2, 9, 17, 21, 24, 28–9, 44, 58, 64, 88, 109
racial inferiority rhetoric, 47–8
racial laws, 110, 115
racial protectionism, 11, 29, 51, 68, 70, 76, 82–3, 87, 98, 107–9, 115
racial purity issues, 47, 99, 101–2, 104, 114
racial resettlement, 87
racial welfare, 113–14
Radi, Lazër, 102
Râmneanțu, Petru, 109
rational selection see artificial selection
Ravasz, László, 87
Reche, Otto, 76
redemption narratives, 101
Reformed Church, 62, 87

regeneration, 5, 10, 21, 25, 29–31, 64, 120–1, 124
  and biopolitics, 100–7
  furthered by war, 42–6
  see also palingenesis; rejuvenation
rejuvenation, 1, 4, 11, 15, 20–2, 26–9, 36–9, 124
  First World War, 40–1, 43, 56, 58, 61
  interwar years, 64, 70–3, 80, 84–6, 93–4, 99, 105–6, 110, 117
Relgis, Eugen, 80
religion, 3–4, 124
  eugenics as, 21–2, 24, 75, 84, 126
  and nationalism, 62–3
  scientism as, 14–15, 18–19, 27
religious support/opposition, 84–91
Rentoul, Robert, 28
Repp, Kevin, 41
reproduction control, 10, 29–31, 41, 70, 83, 115, 124
  see also birth control
reproductive hygiene, 33
responsibility, collective, 8, 58–9, 76, 92, 108
Ribot, Théodule, 28
Ricci, Marcello, 109
Richardson, Angelique, 17
Richer, Ingrid, 88
Richter, Melvin, 3
Robin, Paul, 29–30
Rolland, Romain, 80
Roma, 116
Romania, 30, 74, 78, 96–7, 101, 105, 108–9, 115–16
Romanian eugenicists, 76, 78, 80, 82, 85–6, 99, 104–6, 110, 112
Rüdin, Ernst, 26, 84
Rusev, Ivan, 96
Russian Empire, 45, 107
Růžička, Vladislav, 70, 75

Saleeby, Caleb, 22, 31
Savorgnan, Franco, 48
Saxon Protestant Church, 87
Saxons in Transylvania, 109
Schallmayer, Wilhelm, 25–6, 32, 59
Scholz, Alois, 82
Schorske, Carl E., 118–19
Schreiber, Georges, 98
Schuster, Edgar, 23
Schwartz, Michael, 120
scientism, 1–2, 14–15, 18–19, 21, 27, 29, 101, 116, 118–19, 125
Second World War, 8, 11, 115–16
Section for Social Hygiene and Eugenics, Poland, 74
Segel, Binjamin, 44–5
segregation, 32, 81–2, 84, 114
Sekla, Bohumil, 96
selection, 33, 112–13
  artificial, 26, 33, 38, 42, 44, 64, 79, 94
  natural, 34, 42, 46
Serbia, 30
Sergi, Giuseppe, 25–6, 29, 38, 46, 68
serology, 18
Siegmund, Heinrich, 31
Siemens, Hermann, 70
Sinclair, May, 45
Slovak Republic, 103
social engineering, 5, 12, 21, 68, 77, 112
social hygiene, 8, 23, 32, 35, 65, 77, 84, 118
social palingenesis, 43, 45, 86
social problems, 9, 36–7, 72, 99
social protectionism, 29, 51–3
social welfare, 28, 34, 46, 57, 59, 68, 95
societies *see* associations and societies
Somogyi, József, 89
Sorel, Georges, 29

Spanish eugenicists, 26–7, 30, 41, 85, 96
Spiegel, Gabrielle M., 119
Štampar, Andrija, 69
Stan, Liviu, 85
Stanojević, Vladimir S., 75
state-oriented eugenics, 35–6, 57, 68, 77–8, 83, 110–17, 124
Station for Experimental Evolution, America, 33
Stepan, Nancy, 123
sterilisation, 32, 34–5, 38, 66–7, 70–2, 80–4, 89, 94–6
sterilisation laws, 38, 83, 96–7, 99
stigmatisation, 11, 65–72, 122
Stöcker, Helene, 57–8
Sudeten Germans, 108–9
Sutzu, Alexandru, 30
Swedish eugenicists, 77, 81, 97
Swedish Institute for Race Biology, 74–5
Swedish minority in Finland, 107–8
Swiss eugenicists, 47, 65, 83, 96
Székely, Edmond, 80
Szél, Tivadar, 62–3

Tandler, Julius, 52
Third Greek Civilisation, 103
Thomson, Arthur, 42–3
Tiso, Jozef, 103
Todorov, Tzvetan, 14–15, 120
totalitarianism, 11–12, 64, 98, 112, 116

Tóth, Tihamér, 89
Transylvania, 31, 80, 85, 87, 104, 109
Turkey, 59, 96, 98, 104–5
Turkish eugenicists, 98, 105

Ude, Johann, 88
unfit elements, 28–9, 31–2, 34–5, 38, 46, 66, 72, 79
US, 33, 38, 70–1, 80
utopianism, 15, 22, 35, 61, 91

Vajtauer, Emanuel, 103
Vallejo-Nágera, Antonio, 41, 85
valued elements of society, 32, 42, 46, 64
Viennese Society for Racial Hygiene, 74
vocabulary, eugenic, 11, 22–3, 28, 48, 60, 114, 116
Voegelin, Eric, 14

Wagenen, Bleeker van, 38
war see Boer War; First World War; Second World War
Webb, Sidney, 27
Weindling, Paul, 71
Weiss, Sheila Faith, 25–6, 56
welfare, 28, 34, 46, 57, 59, 68, 95
Wolff, Johannes, 86

Yugoslavia, 75

Zionism, 45
Zurukzoglu, Stavros, 76–7